Molecular Soft-Interface Science

Mizuo Maeda · Atsushi Takahara ·
Hiromi Kitano · Tetsuji Yamaoka ·
Yoshiko Miura
Editors

Molecular Soft-Interface Science

Principles, Molecular Design, Characterization and Application

Springer

Editors
Mizuo Maeda
RIKEN
Wako, Japan

Hiromi Kitano
Toyama University
Toyama, Japan

Yoshiko Miura
Kyushu University
Fukuoka, Japan

Atsushi Takahara
Kyushu University
Fukuoka, Japan

Tetsuji Yamaoka
National Cerebral and Cardiovascular Center
Suita, Japan

ISBN 978-4-431-56875-9 ISBN 978-4-431-56877-3 (eBook)
https://doi.org/10.1007/978-4-431-56877-3

Library of Congress Control Number: 2019935834

This Springer imprint is published by the registered company Springer Japan KK part of Springer Nature
The registered company address is: Shiroyama Trust Tower, 4-3-1 Toranomon, Minato-ku, Tokyo 105-6005, Japan

Preface

This book is aimed as an introductory textbook for graduate students and researchers interesting in the study of soft material surface and interfaces, including industrial applications. The book provides comprehensive basic knowledge and applications related to soft interfaces.

A "soft interface" refers to any shared boundary between a bulk phase (liquid, air or solid) and a soft material, which are flexible, organic materials, such as polymers, gels, amphiphiles, colloids and liquid crystals, that may be responsive to external stimuli. The soft interface itself is a boundary area with three-dimensional structure and depending on the behavior of the organic material, may have tailorable response that is sensitive to temperature, ions, solvents and guest molecules. The properties of soft interfaces are governed by the structure and dynamics of the soft material, which may depend on self-assembly and supramolecular structure.

In addition, since soft materials also include biopolymers such as proteins, nucleic acids and polysaccharides, the study of soft interfaces is also strongly related to the fields of biomembranes and biomaterials. Soft interfaces with biomembranes must have suitable properties to maintain biological activity, and will determine the functionality of biomaterials and bio-devices.

Soft interfaces are currently one of the most important research areas in the development of functional materials. Studies of soft interfaces are by necessity interdisciplinary, and require background in synthetic chemistry, physical science, materials science and analytical science. Soft interfaces are also of significant interest in the fields of chemical, mechanical, biological and medical engineering. Due to the highly multidisciplinary scope, soft interfaces have not been categorized into any specific, existing research academic discipline. A comprehensive knowledge of soft interface has not yet been systematically presented in the area of chemistry, physics and materials science.

The primary aim of this book is to present a detailed overview of the scope of interface science, preparation of soft materials and soft interfaces, analyses of soft interfaces and application to functional materials. Four editors and fifteen authors have contributed. In Part I, the science of soft interfaced will be described from the viewpoint of physical chemistry. In Part II, methods of materials preparation will be described, including polymer chemistry, polymer brush preparation and supramolecular chemistry. Part III will provide an introduction to the analysis of soft interfaces. In Part IV, various applications of soft interfaces will be presented, with detailed focus on the application of biomaterials. As necessary, each chapter is able to be downloaded separately based on the index. We hope this book provides the desired and useful knowledge to the students and researcher in both academia and industry.

Fukuoka, Japan Yoshiko Miura

Contents

Part I
The Principle and Physical Chemistry of Soft Interface

Chapter 1
The Principle and Physical Chemistry of Soft Interface

Takanori Takiue, Yoshimune Nonomura and Syuji Fujii

1.1 Colloid and Interface (Molecular Force, Colloid, Air/Water Interface)

1.1.1 Molecular Force

Intermolecular forces are classified into three categories; (1) purely electrostatic Coulomb force, (2) polarization force, and (3) quantum mechanical force. The interaction between charges, permanent dipoles, etc. belongs to the first category. The polarization force is the interaction arises from the dipole moments induced in atoms and molecules by the electric field of charges and permanent dipoles. The covalent or chemical bonding is categorized into the quantum mechanical force. The first and second forces show an order of ~kT per mole and play a crucial role to determine their structures and properties of colloid and interface [1, 2].

The interaction energy between two charges Q_1 and Q_2 is given by

$$w = \frac{Q_1 Q_2}{4\pi\varepsilon_0 \varepsilon r^2} \tag{1.1}$$

T. Takiue
Faculty of Arts and Science, Kyushu University, 744 Motooka, Nishi-ku, Fukuoka 819-0395, Japan
e-mail: t.takiue@chem.kyushu-univ.jp

Y. Nonomura
Department of Biochemical Engineering, Graduate School of Science and Engineering, Yamagata University, 4-3-16, Jonan, Yonezawa, Yamagata 992-8510, Japan
e-mail: nonoy@yz.yamagata-u.ac.jp

S. Fujii (✉)
Department of Applied Chemistry, Faculty of Engineering, Osaka Institute of Technology, 5-16-1, Omiya, Asahi-Ku, Osaka 535-8585, Japan
e-mail: syuji.fujii@oit.ac.jp

M. Maeda et al. (eds.), *Molecular Soft-Interface Science*,
https://doi.org/10.1007/978-4-431-56877-3_1

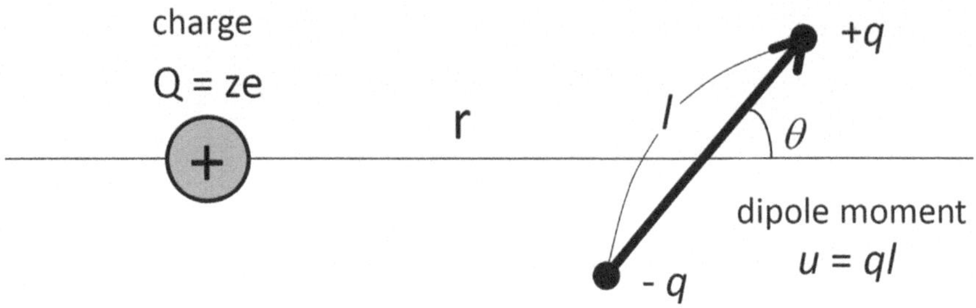

Fig. 1.1 A charge Q at a distance from the center of a polar molecule of dipole moment *u* subtending an angle θ to the line joining the two molecules

where ε is the relative dielectric constant of medium and r is the distance between two charges. This interaction is attractive for like charges and repulsive for unlike charges. In case of two ions (e.g., Na^+ and Cl^-) in contact, $w = -8.4 \times 10^{-21}$ J, which is equivalent to around 200 kT, similar to the covalent bond energy. The ion–dipole interaction such as that between Na^+ ion and water molecule is expressed as

$$w = \frac{\mathrm{Qu\cos\theta}}{4\pi\varepsilon_0\varepsilon \mathrm{r}^2} \tag{1.2}$$

where u is the dipole moment of polar molecule and θ is the orientation angle of the dipole (Fig. 1.1). This interaction is much stronger than kT, and thus strong enough to bind ions to polar molecules and mutually orient with each other. The angle-averaged interaction energy for the charge–dipole interaction is

$$w = -\frac{\mathrm{Q^2u^2}}{6(4\pi\varepsilon_0\varepsilon)^2\mathrm{kTr^4}}, \tag{1.3}$$

which is attractive and temperature dependent. When the two polar molecules are close to each other, the angle-averaged dipole–dipole interaction between them is given by

$$w = -\frac{\mathrm{u_1}^2\mathrm{u_2}^2}{3(4\pi\varepsilon_0\varepsilon)^2\mathrm{kTr^6}}, \tag{1.4}$$

where u_1 and u_2 are the dipole moment of two polar molecules, respectively. This is referred to as Keesom interaction, which is one of three interactions varying with inverse sixth power of distance and contributes to the van der Waals interaction between molecules.

A polar molecule induces polarization of nonpolar molecule and can interact attractively with each other. This dipole–induced dipole interaction is given by

$$w = -\frac{u^2\alpha}{(4\pi\varepsilon_0\varepsilon)^2 r^6}, \tag{1.5}$$

and referred to as Debye interaction, which is also one of the inverse sixth power contribution to van der Waals interaction. Here a is the electric polarizability of molecule. The third contribution is the dispersion interaction which is basically quantum mechanical in nature because it involves interaction between fluctuating dipole induced by the movement of outer valence shell electrons of atom. This force is always present between atoms and molecules, and plays an important role in phenomena such as surface tension, wetting, adhesion, flocculation of colloidal particles, and structure formation of polymers and proteins. The expression for the dispersion interaction between two identical molecules is written as

$$w = -\frac{3}{4}\frac{\alpha^2 I}{(4\pi\varepsilon_0\varepsilon)^2 r^6}, \tag{1.6}$$

where I is the ionization potential.

1.1.2 Colloid

A colloid is s system consisting of dispersed phase (solid, liquid, or gas) divided and distributed in a continuous phase or dispersed medium (solid, liquid, or gas). It is well known that colloidal properties are usually exhibited by systems with the size of dispersed phase of around a few to thousand nm. We usually encounter many colloids in our life; milk, butter, smoke, fog, paints, gels, and so on. The types of dispersed systems are classified depending on the nature of the dispersed and continuous phases. For example, smoke is classified into "aerosol" in which solid particles are dispersed in gas phase. Milk is an emulsion in which liquid oil is dispersed in liquid water. The other class of colloids in the context of surface chemistry is "association colloids", which consist of aggregates or units of a lot of molecules. The formation of association colloids often depends on factors such as temperature, pressure, concentration of substances, and the chemical structure of molecules. Many biological systems, including cell membrane and transport phenomena, involve various forms of colloidal structures.

1.1.3 Stability of Colloid

In Sect. 1.1.1, the interaction between two isolated atoms or molecules was referred. When we consider the long-range interactions between macroscopic particles and surfaces in liquid, we should take into account of three important interactions: van der Waals attraction, electrostatic repulsion, and the interaction at very short distance such as steric forces. They play a crucial role for stabilization of colloid particles in continuous medium [1, 2].

The mutual interaction between two nonpolar molecules depends on the distance of separation of the molecules r. If it is assumed that the interaction energy at infinite separation is zero, the free energy of attraction ΔG_{att} at distance r is written as

$$\Delta G_{\mathrm{att}} = -\frac{A}{\mathrm{r}^6}. \tag{1.7}$$

The constant A for two identical molecules is given by

$$A = \frac{3}{4} h \nu \alpha^2, \tag{1.8}$$

where h is Planck's constant, α is the electric polarizability of atom or molecule, and ν is a characteristic frequency corresponding to the ionization energy. For different molecules 1 and 2, the constant A is given by

$$A_{12} = \frac{3}{4} h \left(\frac{\nu_1 \nu_2}{\nu_1 + \nu_2} \right) \alpha_1 \alpha_2. \tag{1.9}$$

Thus, the attractive force between two molecules increases continuously with decreasing their separation. On the other hand, the electron clouds of two molecules begin to interact and overlap at some distance. If the covalent bonding between the molecules is not possible, it produces repulsion and increases the free energy. This is Born repulsion ΔG_{rep} which is expressed by

$$\Delta G_{\mathrm{rep}} = \frac{B}{\mathrm{r}^{12}}. \tag{1.10}$$

Thus, the total free energy is the sum of the attractive and repulsive contributions

$$\Delta G = \Delta G_{\mathrm{rep}} + \Delta G_{\mathrm{att}} = \frac{B}{\mathrm{r}^{12}} - \frac{A}{\mathrm{r}^6}, \tag{1.11}$$

which is usually known as Lennard–Jones (L-J) 6-12 potential.

Here, let us consider the interaction between colloid particles. In this case, it is assumed that the particles interact mutually according to L-J potential and that the total interaction is the sum of all individual molecular interactions. The repulsive contribution is neglected in a particle and taken into account between the opposing surface of the particles. One of the simplest situations to analyze such interaction is that between two flat surfaces separated by a distance H in vacuum (Fig. 1.2). The free energy of attraction per unit area is given by

$$\Delta G_{\mathrm{att}} = \frac{A_{\mathrm{H}}}{12\pi \mathrm{H}^2}, \tag{1.12}$$

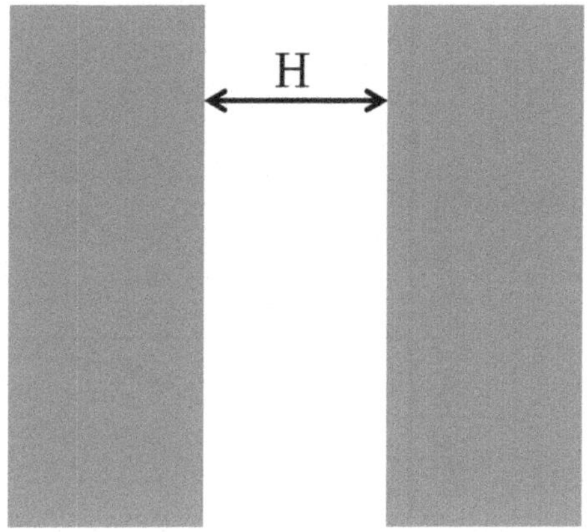

Fig. 1.2 Two interacting planer surfaces at a distance H

where A_H is the Hamaker constant and related to A in Eq. 1.8 by

$$A_\mathrm{H} = \frac{3}{4} h\nu\alpha^2\pi^2 n^2 = A\pi^2 n^2, \tag{1.13}$$

where n is the number of molecules in unit volume of the phase. In case of the interaction between two identical spheres of radius a, where $\mathrm{H/a} \ll 1$, the expression becomes to be

$$\Delta G_\mathrm{att} = -\frac{A_\mathrm{H}\alpha}{12\pi\mathrm{H}}\left[1 + \frac{3}{4}\frac{\mathrm{H}}{\mathrm{a}} + \cdots\right]. \tag{1.14}$$

Equations 1.13 and 1.14 indicate that the free energy of attraction between two surfaces falls off much slowly than that between molecules. This extended range of surface interaction plays an important role in determining the properties of colloidal systems.

When two surfaces (component 1) interact through continuous medium (component 2), the effective Hamaker constant, A'_H, is given by

$$A'_\mathrm{H} = \left[\sqrt{A_{10}} - \sqrt{A_{20}}\right]^2, \tag{1.15}$$

where A_{10} and A_{20} are respectively the Hamaker constant of component 1 and 2 in vacuum.

The repulsive interaction between two colloidal particles separated by a distance H arises from the overlap of electrical double layers of them. The expression for this is approximated as

$$\Delta G_\mathrm{rep} = \frac{64 c_0 k Y \gamma^2}{\kappa} \exp(-\kappa\mathrm{H}), \tag{1.16}$$

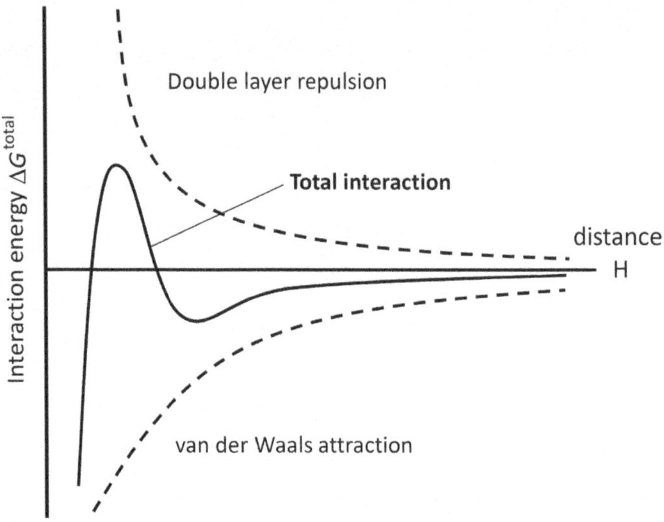

Fig. 1.3 Schematic profile of interaction energy versus distance between particles

where c_0 is the concentration far from the surface of particle, $1/\kappa$ corresponds to the thickness of electrical double layer, and γ is given by

$$\gamma = \frac{\exp(\mathrm{ze}\psi_s/2kT) - 1}{\exp(\mathrm{ze}\psi_s/2kT) + 1}. \tag{1.17}$$

Here, ψ_s is surface potential of the particle, z is the charge of counter ion, and e is the unit electrical charge.

Thus, for planar surface, the total interaction $\Delta G^{\mathrm{total}}$ is the sum of ΔG_{att} and ΔG_{rep}, and expresses as

$$\Delta G^{\mathrm{total}} = \frac{64c_0kT\gamma^2}{\kappa}\exp(-\kappa \mathrm{H}) - \frac{A'_{\mathrm{H}}}{12\pi \mathrm{H}^2}. \tag{1.18}$$

A typical interaction energy curve is illustrated in Fig. 1.3. It is noted that the height of energy barrier depends on the concentration and valence of electrolyte; an increase in electrolyte concentration reduces the repulsive interaction and eventually reduces the energy barrier. This promotes collision and coagulate colloidal particles and thus the system is less stable.

1.1.4 Surface Adsorption

One of the most important topics in surface and colloid science is adsorption of molecules at interfaces. In particular, the adsorbed film of surface active substances at soft interfaces including gas/liquid and liquid/liquid interfaces is a fundamental structure of more complicated molecular organizing systems, soft matter, such as emulsion, vesicle, biological membrane, and so on. Thus the study on the structure and property of soft interface is indispensable to understand accurately the structure–function relation of soft matters. In this section, we will introduce recent developments in the study on the adsorbed films at gas/liquid interface from molecular level by means of X-ray, laser beams, etc., as well as from a macroscopic viewpoint based on the surface tension measurement.

The pure gas/liquid interfaces as well as the adsorbed films provide fundamental information on the structure and property of interfacial films. Figure 1.4 shows the surface tension γ versus temperature T curve measured for a pure *n*-octadecane (C18)/air surface. The curve has a distinct breakpoint, below which the γ value decreases steeply with decreasing T, indicating a well-ordered structure at the surface. Thus, this phenomenon is called "surface freezing (SF)". The surface freezing phenomenon was found in the liquid alkane with carbon number of $16 \leq n \leq 50$ and alkanol with even carbon number of $10 \leq n \leq 28$. The SF layer formation is clearly seen by the appearance of modulations in the X-ray reflectivity curve (Fig. 1.5). According to the electron density profile of alkane, alkane molecules form condensed monolayer in which the molecules are closely packed like solid rotator phase. Furthermore, the in-plane order examined by grazing incidence X-ray diffraction (GIXD) confirmed the hexagonal molecular arrangement with almost perpendicular orientation [3–5].

The state of adsorbed film of single-component system is controlled by temperature, pressure, and concentration of the surfactant solution. Especially, the structure of adsorbed film at air/water interface can be characterized by a lot of sophisticated techniques such as Brewster angle microscopy (BAM) [6, 7], external reflection-absorption FTIR spectroscopy (ERA FTIR) [8], ellipsometry [9], XR [10, 11], GIXD [11, 12], and neutron reflection (NR) [13, 14]. The morphology of condensed domains of 1-dodecanol (C12OH) at the aqueous solution surface studied by BAM coupled with dynamic surface tension measurement indicated that the condensed domains are formed after the breakpoint on the γ versus time curve. The domains grow gradually with time and finally approach a homogeneous film close to the adsorption equilibrium (Fig. 1.6) [15].

The adsorbed film of ionic surfactant, decyltrimethylammonium bromide (C10TAB) at the air/water interface was examined by NR. The quantitative analysis of the reflectivity using optical matrix method provides a structural parameter. At 0.002 M, the thickness of the film is 15 Å, indicating that the molecules stand almost upright at the interface. On the other hand, the thickness was found to be around 21 Å at high concentration where the monolayer is essentially saturated. This suggests that as the charged head groups (timethylammonium ion) are forced into

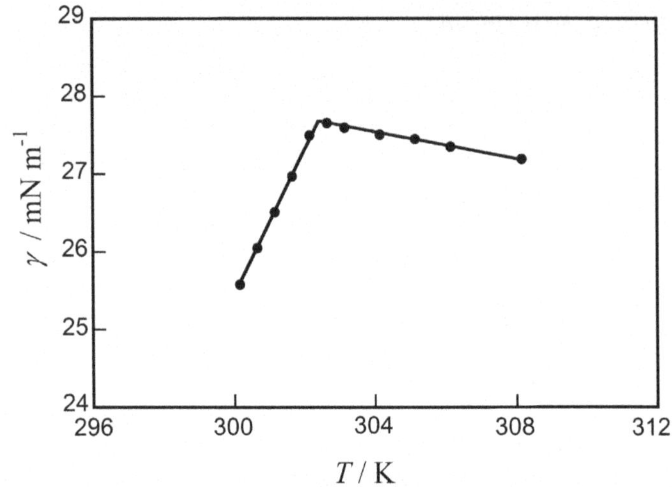

Fig. 1.4 Surface tension versus temperature curve of liquid octadecane

Fig. 1.5 X-ray reflectivity versus scattering vector from liquid alkane surfaces

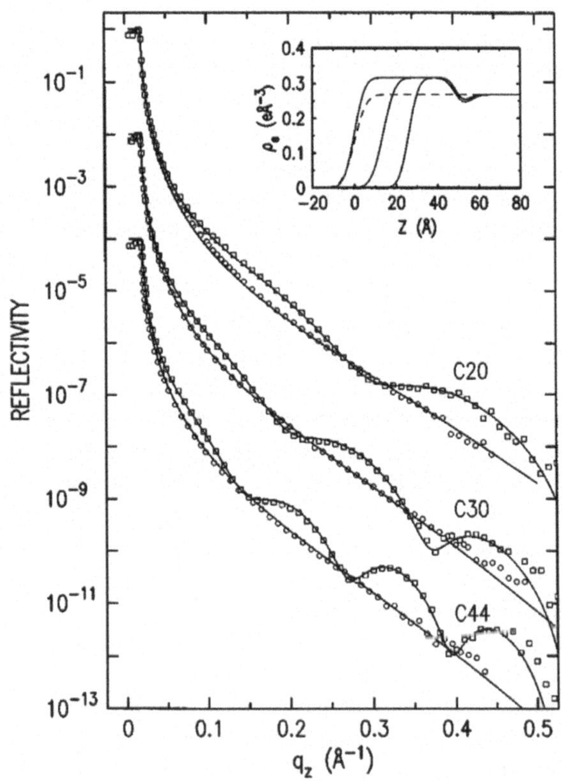

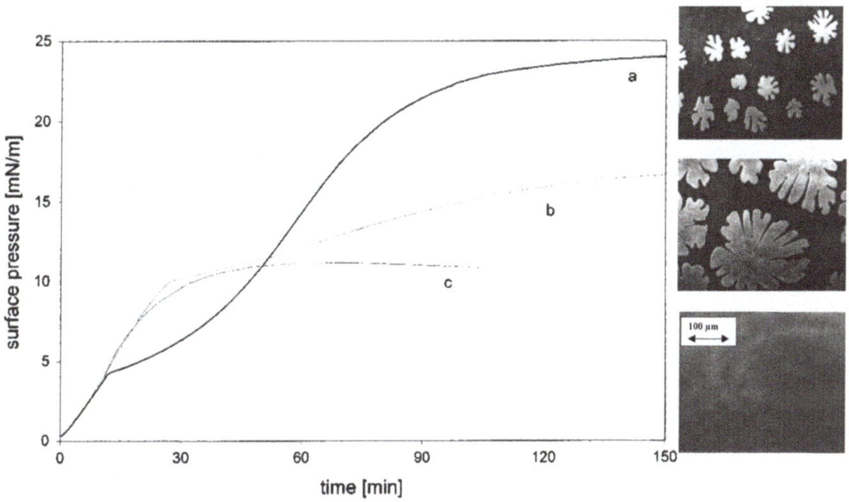

Fig. 1.6 Dynamic surface pressure measurement of aqueous 1-dodecanol solution surface coupled with BAM observation

Fig. 1.7 Staggered arrangement of adsorbed surfactant ions at air/water surface

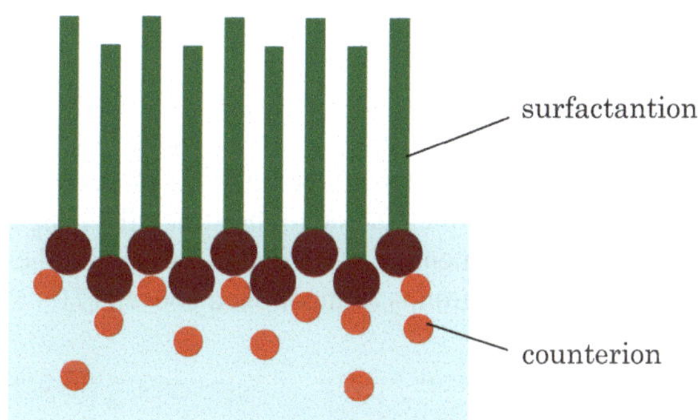

a smaller area with increasing concentration, their mutual repulsive force induces them to take a staggered arrangement as illustrated in Fig. 1.7 [16].

The counterions distribute around the polar head group with an opposite charge in the adsorbed film of ionic surfactant. The X-ray absorption fine structure (XAFS) technique was applied to the aqueous dodecyltrimethylammonium bromide (C12TAB) solution surface under total reflection condition in order to estimate directly the surface concentration of Br ions and consider the structure of adsorbed film from the viewpoint of counterion distribution in the interfacial region. Figure 1.8 shows some typical XAFS spectra obtained at different bulk concentrations *m*. The K-edge absorption jump (*J* value) increases with increasing concentration. The shape of the spectra also varies with concentration; the shift of position of the first maximum and minimum was observed. This suggests the change in hydration structure of Br ion with concentration. The *J* values from XAFS measurement are plotted

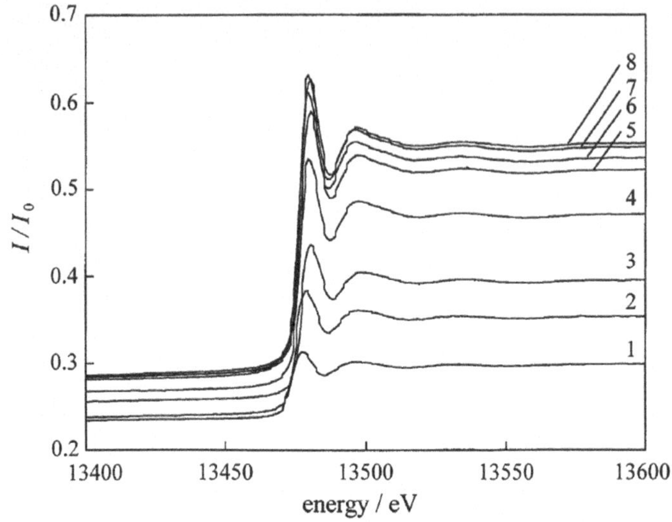

Fig. 1.8 Total reflection XAFS spectrum at the C12TAB solution surface at given molality; (1) 1.00, (2) 1.50, (3) 2.00, (4) 5.00, (5) 10.0, (6) 15.0 (7) 22.5, and (8) 25.0 mmol kg^{-1}

and compared to the surface density Γ^{H} calculated from the surface tension versus concentration curve in Fig. 1.9. The J values trace almost perfectly the Γ^{H} versus m curve. Furthermore, the J value starts to increase close to the critical micelle concentration (cmc), which is due to the structure change of adsorbed film, such as the staggered arrangement of DTA ions to minimize the electric repulsive interaction between the head groups as mentioned above [17].

1.2 Wettability and Molecular Science

1.2.1 Surface Tension

Surface tension is the excess energy that arises from the presence of a surface. This excess energy induces some characteristic geometrical structures such as the complete spherical shape of a soap bubble and a raised water surface at the edge of a cup. The origin of the energy is an imbalance of intermolecular forces at surfaces. Figure 1.10 shows the location of water molecules at an air–water interface. The cohesion energy, which evolved from hydrogen bonds and van der Waals attractions between water molecules, decreases the energy and stabilizes the system. The presence of a surface increases the energy because the molecules at the surface do not have other adjacent water molecules for the formation of intermolecular interactions. Surface tension is defined as an increased energy per unit area. The following are data

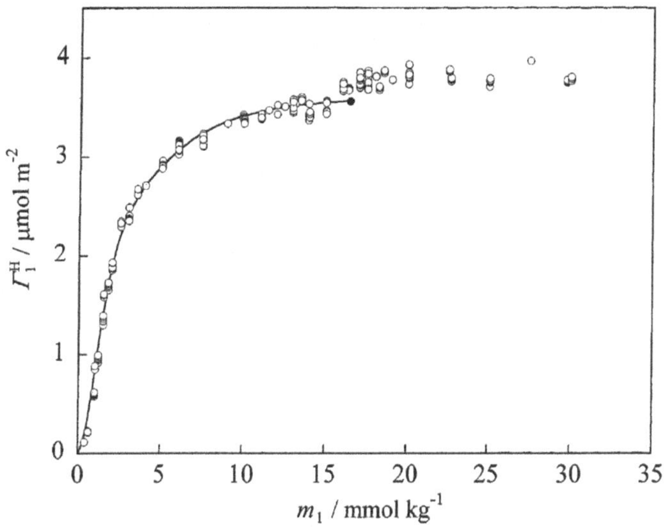

Fig. 1.9 Surface concentration and K-edge jump J value versus molality curve of aqueous C12TAB solution surface; (—)surface tension method, (○) XAFS method

of surface tensions at 293 K for some liquids: water (72.8 mN m^{-1}), toluene (28.4 mN m^{-1}), chloroform (27.1 mN m^{-1}), diethyl ether (17.0 mN m^{-1}), and mercury (476 mN m^{-1}). In general, the surface tension of solid materials is larger than that of liquid materials because intermolecular interactions between molecules in solid materials generate greater cohesion energy than in liquid materials. However, the surface tension of polymers is often small when they are in solid state having small cohesion energy per unit volume. These solid systems are prevented from degradation by thermal energy owing to their huge size. The following are surface tensions of some metals and polymers in solid state: gold (1205 mN m^{-1}, 973 K), silver (1140 mN m^{-1}, 1173 K), iron (1670 mN m^{-1}, 1673 K), polyethylene (34–37 mN m^{-1}, 293 K), and tetrafluoroethylene (22–24 mN m^{-1}, 293 K).

Excess energy, referred to as interfacial tension γ_{OW}, arises at oil/water interfaces. When two surfaces of water and oil exist independently, these surfaces give rise to excess energies with surfaces tensions γ_{W} and γ_{O}, respectively. In the case of an oil/water interface, the excess energies originated from oil and water surfaces are eliminated by attractive interactions between water and oil molecules at the interfaces. If σ_{OW} is the cohesion energy between water and oil molecules, the interfacial tension γ_{OW} can be described by Eq. 1.19.

$$\gamma_{\mathrm{OW}} = \gamma_{\mathrm{O}} + \gamma_{\mathrm{W}} - 2\sigma_{\mathrm{OW}}. \tag{1.19}$$

The interfacial tensions between some oils and water are as follows: toluene (36 mN m^{-1}, 298 K), chloroform (31.6 mN m^{-1}), and diethyl ether (10.7 mN m^{-1}). The

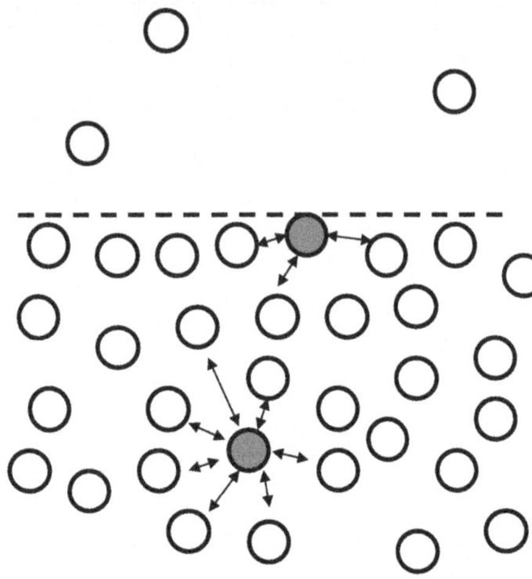

Fig. 1.10 Location of water molecules at an air–water interface

addition of surfactant decreases interfacial tension because the surfactant molecules are adsorbed at the oil/water interface and decrease the interfacial tension. When σ_1 and σ_2 are the cohesion energies between a lipophilic group in a surfactant molecule and an oil molecule, and between a hydrophilic group and a water molecule, respectively, the interfacial energy adsorbing surfactant molecules can be described by Eq. 1.20.

$$\gamma_{\mathrm{OW}} = \gamma_{\mathrm{O}} + \gamma_{\mathrm{W}} - (\sigma_1 + \sigma_2) \tag{1.20}$$

In general, σ_2 is larger than σ_{OW}, while σ_1 is similar to σ_{OW}. This is the reason why surfactant molecules decrease interfacial tension between oil/water interfaces.

1.2.2 *Wettability*

Wetting is an interfacial phenomenon by which a liquid droplet spreads on a solid surface. This is a concern with many industrial products such as detergents, automobiles, foods, cosmetics, and biological phenomena in animal bodies and on plant surfaces. Characteristic interfacial phenomena have been observed during superhydrophobic and self-cleaning phenomena of lotus leaves, rose petals, and the compound eyes of mosquitoes [18–20]. This also observed with the wetting on the hydrophilic mucosal membrane of eyes, the small intestinal wall [21], and the floating of water striders [22]. As shown in Fig. 1.11, the contact angle of a liquid droplet conforms to Young's law on a homogeneous flat surface. The interfacial tensions between liquid, gas, and solid phases are balanced at a three-phase contact line [23]: the contact angle at equilibrium state θ_{E} is

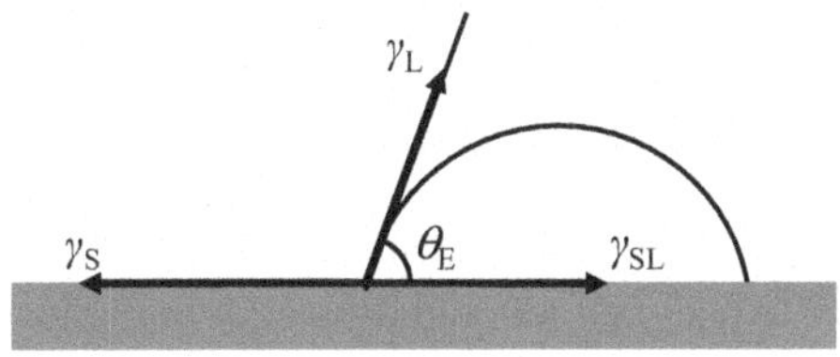

Fig. 1.11 Contact angle of a liquid droplet on a solid surface

$$\cos\theta_E = \frac{\gamma_S - \gamma_{SL}}{\gamma_L}, \tag{1.21}$$

where γ_S, γ_L, and γ_{SL} are surface tensions of the solid, liquid, and interfacial tension between solid and liquid, respectively.

Roughness of solid surfaces affects the wetting phenomena because rough structures increase actual surface area and enhance the surface properties of solid materials. The contact angle conforms to Wenzel's law on a solid surface with fine textures when the liquid completely comes in contact with the solid surface [24]. When the surface area is increased R times with a rough structure, the surface tension of the solid surface and the interfacial tension between solid and liquid in Eq. 1.21 are multiplied by R. Then, the contact angle on rough surfaces θ_E^* can be described as follows:

$$\cos\theta_E^* = R\frac{\gamma_S - \gamma_{SL}}{\gamma_L} = R\cos\theta_E \tag{1.22}$$

Here if cos θ_E is positive ($\theta_E < 90°$) or negative ($\theta_E > 90°$), $\cos\theta_E^*$ is a larger or smaller value, respectively. This implies that if a solid surface is roughened, the hydrophilic (hydrophobic) surface becomes more hydrophilic (hydrophobic).

When depressions on a rough structure are deeper than a critical length, the contact angle conforms to the Cassie–Baxter law that states that a gas phase exists between water and solid surfaces [25]. In this model, the solid surface is assumed to be a composite surface consisting of two materials. The contact angle θ_E^* can be described as follows in the equilibrium state:

$$\cos\theta_E^* = f_1\cos\theta_1 + f_2\cos\theta_2, \tag{1.23}$$

where θ_1 and θ_2 are the contact angles on flat and homogeneous surfaces of materials 1 and 2, respectively. f_1 and f_2 are area fractions of materials 1 and 2 on the solid surface, respectively. If the water does not reach to the bottom of the depression, the second component can be assumed to be air. When the contact angle between water and air is 180°,

$$\cos\theta_E^* = f - 1 + f\cos\theta_E. \tag{1.24}$$

This predicts that the right side approaches −1 when the solid fraction f decreases without limit. This implies that the contact angle of water on a pillar surface covered with ultrafine rods is almost 180°.

As shown above, we can describe the morphology of a water droplet on a solid surface based on some well-established theoretical models. In the equilibrium state, we expect that a liquid droplet is in the lower energy state of the Wenzel state and Cassie–Baxter state. However, these models do not adequately describe the wetting phenomena for real-life situations because the systems consisting of a liquid droplet and a solid surface can be under another metastable state because of an energy barrier between Cassie–Baxter and Wenzel states. For example, a metastable Cassie–Baxter state changed to a Wenzel state as pressure was applied from above [26]. Some theoretical studies proposed an intermediate state at the conversion process between two states [27].

A dynamic process has also been studied for the spreading behavior of liquid materials. When a liquid droplet is dropped on a solid surface, the contact angle θ_{D} is proportional to power-of-time t.

$$\theta_{\mathrm{D}} \sim t^{-3/10} \tag{1.25}$$

This rule is called Tanner's law and is applied when a nonvolatile liquid is spread on flat and clean surfaces. The multiplier drastically increases if the solid surfaces are covered by rough structures or if the liquid phase contains volatile material, which induces Marangoni flow [28, 29].

Artificial materials with characteristic wettability have been developed over several decades. The typical materials are superhydrophobic materials, which repel water. The surface of a wax crystal, alkylketendimer, exhibits superhydrophobicity owing to its hierarchical rough structure [30]. The contact angle between a water droplet and its surface was 174°. Similar hydrophobic phenomena were observed on water-repellent multipillar surfaces [31], Teflon-coated carbon nanotube forests [32], polymer honeycomb surfaces [33], and moss eye-mimetic surfaces [20]. Moreover, super oil-repellent surfaces [34, 35], superhydrophilic surfaces [36], and switching material have been developed and applied in many fields.

1.3 Surfactants (Structure and Function, Emulsion)

1.3.1 Surfactants

What is a surfactant?

Amphiphilic molecules are used in everyday life and in many industrial processes. For example, they are used as soaps, detergents, wetting agents, dispersants, emulsifiers, foaming agents, bactericides, corrosion inhibitors, antistatic agents, and flotation agents. They are also used to form membranes, polymer particles, liposomes, vesicles and (micro)gels. A media-soluble amphiphilic molecule that is surface active is called a surfactant, which is a contraction of the phrase "surface active agent" (note that "amphi" means "of both kinds" in Greek). Surfactants are substances that can adsorb onto surfaces/interfaces at low concentrations in a system and alter the surface/interfacial free energies to a marked degree. Surfactants have hydrophilic (compatible with water) and hydrophobic (non/poorly compatible with water) components in their structures and adsorb strongly at interfaces such as oil–water and air–water. The hydrophobic part is sometimes called the lipophilic (compatible with fat and oil) part. The hydrophobic part can be a hydrocarbon, a fluorocarbon or a siloxane. For a surfactant dissolved in aqueous medium, the hydrophobic part distorts the structure of water by breaking hydrogen bonds between the water molecules and by restructuring the water in the vicinity of the hydrophobic group. As a result of this distortion, some of the surfactant molecules are expelled to the interfaces of the system, with their hydrophobic groups oriented in order to minimize contact with the water molecules. There are numerous books and reviews on surfactants [37–41].

Structure

Surfactants are classified into four groups depending on variation of charge, namely anionic, cationic, nonionic, and zwitterionic (Table 1.1).

Anionic surfactants carry negatively charged hydrophilic groups such as carboxylates, sulfonates, or sulfates. Sodium dodecylsulfate (SDS) is one of the most important and widely used anionic surfactants. In water, the alkali metal dissociates as a cation and the surfactant becomes negatively charged.

Cationic surfactants have positively charged hydrophilic groups. For example, cetyl ammonium bromide (CTAB) dissociates in water with the positive charge localized at the quaternized nitrogen. Cationic surfactants can adsorb strongly onto most solid surfaces, which are usually negatively charged. Generally, cationic surfactants are more expensive than anionic or nonionic surfactants.

Nonionic surfactants carry no charge. The hydrophilic part consists of poly(ethylene oxide) or sugars, which are neutral and water soluble. Nonionic surfactants can be divided into alkylethylene oxides, alkylethylene glycols, and alkyl glycosides. They can adsorb onto surfaces with either their hydrophilic or hydrophobic parts, depending on the nature of the surface.

Zwitterionic surfactants carry both positive and negative charges with a net charge of zero. Examples are synthetic products like betaines or sulfobetaines and natural substances such as lipids (e.g., phosphatidylcholine). They can adsorb onto both negatively charged and positively charged surfaces without changing the surface charge significantly.

Surfactants can be also classified depending on their structure. Conventional surfactants have one polar and one nonopolar part. In addition to these surfactants, dimeric (Gemini-type) and oligomeric surfactants have attracted great interest in academia and industry. Currently, there is increasing interest in polymeric surfactants including polymers where each monomer unit is amphiphilic, and block polymers and graft polymers carrying hydrophilic and hydrophobic polymer units [42].

Table 1.1 Chemical structure and substance name of surfactants and critical micelle concentrations (CMC) in water at 25°C (no added salt)

Chemical structure	Substance name	CMC/mM
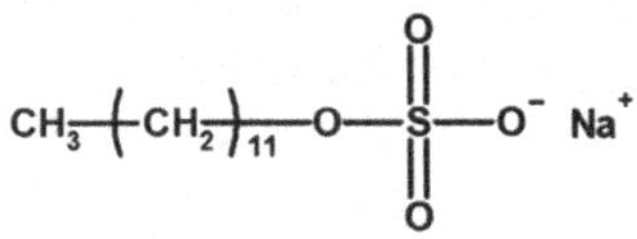	Sodium dodecylsulfate (SDS)	8.9
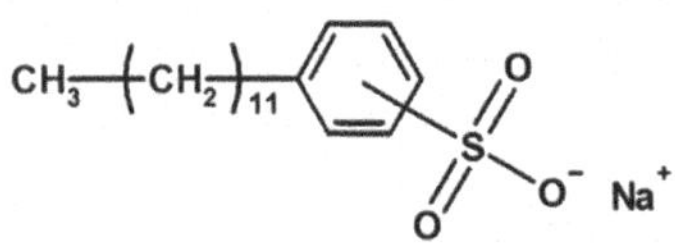	Sodium dodecylbenzene sulfonate	3.6
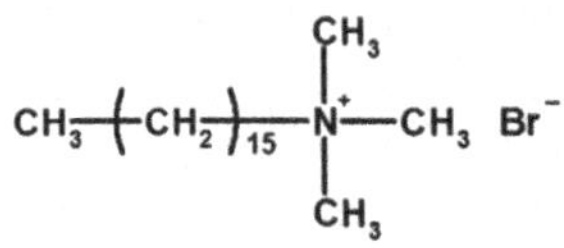	Hexadecyl trimethylammonium bromide (CTAB)	0.9
$C_{12}H_{25}(OCH_2CH_2)_8OH$	Octaethylene glycol monododecyl ether	0.071
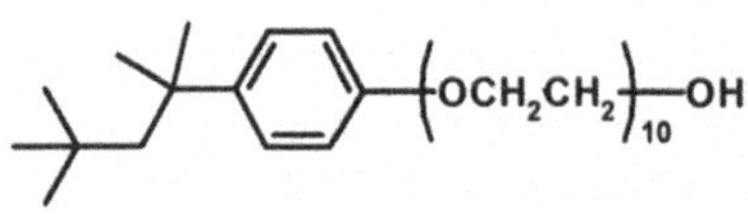	Poly(ethylene oxide) *iso*-octylphenyl ether (Triton® X-100)	0.24
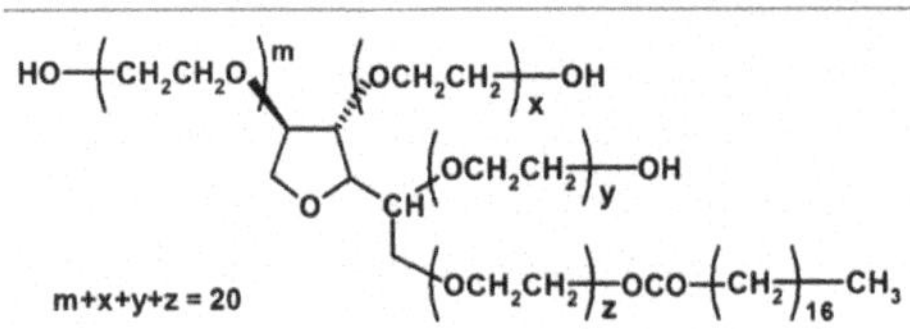	Poly(ethylene oxide) sorbitan monostearate (Tween®60)	0.0027
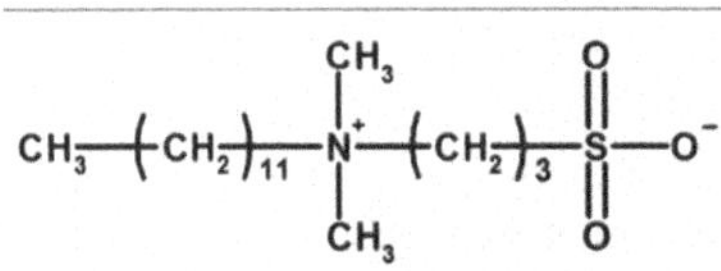	*N*-dodecyl-*N*,*N*-dimethyl propanesultaine	–

Poly(ethylene oxide)-poly(propylene oxide)-poly(ethylene oxide) triblock polymer, known as Pluronics, is one of these commercially available polymeric surfactants.

Surfactants in solution
Influence of temperature

There are two characteristic temperatures in surfactant aqueous solutions. The Krafft temperature (Krafft point) is the point at which the ionic surfactant solubility in water increases drastically and micelles start to form when the temperature increases. Below the Krafft point, the solubility of the surfactant is quite low and the surfactant precipitates as hydrated crystals or as a liquid crystalline phase. Nonionic surfactants tend to show the opposite temperature effect. As the temperature is raised, large aggregates precipitate out into a distinct phase at a temperature called the cloud point.

Critical micelle concentration

Spontaneous aggregation of surfactants in water is an important phenomenon because a number of important interfacial (e.g., detergency and solubilization) and bulk (e.g., viscosity) properties depend on the existence of aggregation. The simplest and best understanding of the aggregation can be explained with spherical micelles. Aggregation occurs when more and more surfactant is added to water. At lower concentration, the surfactant molecules are dissolved as individual ions. Due to their hydrocarbon chains, the molecules tend to adsorb at the air–water interface, with their hydrophobic chains oriented toward the air phase. With increasing surfactant concentration, the surface tension decreases greatly. At a certain concentration, the decrease in surface tension stops. The concentration at which this occurs is called the critical micelle concentration (CMC). For example, the CMC for SDS in water is 8.3 mM. Above the CMC, the surface tension remains almost constant. Similar dependencies on surfactant concentration are observed for osmotic pressure, electrical conductance, or turbidity. The solubility of hydrophobic molecules changes drastically below and above the CMC. Below the CMC, hydrophobic substances are poorly dissolved in aqueous media, but at and above the CMC they become soluble. The micelle has a structure in which the hydrophobic part gathers inside the aggregate and the hydrophilic part orients toward the aqueous phase. Micelles typically consist of 30–100 surfactant molecules and have diameters of 3–6 nm, as determined by light scattering, small-angle X-ray scattering and small-angle neutron scattering measurements (Fig. 1.12). The CMC can be determined by measuring the changes in physical properties such as electrical conductivity, surface tension, light scattering intensity, or solubility of hydrophobic fluorescent substrates.

Micelles are formed because of two competing factors [43]. The first factor is the hydrophobic effect, i.e., the transfer of hydrophobic parts out of water into the oil-like interior drives micellization. This is an entropic effect because highly ordered water molecules around the hydrophobic parts can be released into aqueous media by micellization for entropy to increase. The second factor is repulsion between the hydrophilic parts. Charged hydrophilic parts repel each other electrostatically and steric repulsion works for nonionic hydrophilic parts.

Structure of surfactant aggregates

In addition to the spherical micelles discussed above, surfactants can form cylinders (rod-shaped micelles with hemispherical ends), bilayers (flat lamellar disk-shaped micelles), vesicles (more or less spherical structures consisting of bilayer lamellar micelles arranged in one or more concentric spheres), or inverted micelles (Fig. 1.13). The surfactant parameter, also known as the critical packing parameter, can be used as a guide to the aggregate architecture for a given surfactant:

$$V_C/L_C\sigma_A$$

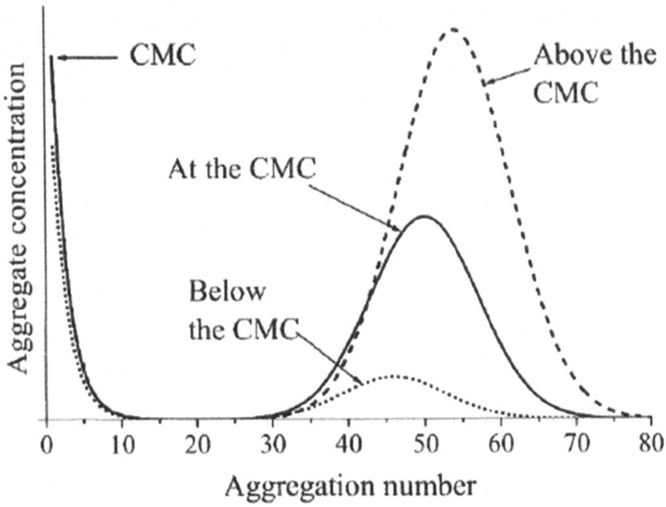

Fig. 1.12 Schematic distribution of surfactant aggregates as a function of the aggregation number for three different concentrations. A mean aggregation number of 50 can be determined at the concentration where the total surfactant concentration is equal to the CMC. (Reproduced with permission from Ref. [37])

where V_C is the volume of the hydrophobic part of the surfactant, L_C is the length of the hydrophobic chains, and σ_A is the effective area per hydrophilic head group. Typical values and their corresponding aggregate structures are:

$V_C/L_C\sigma_A$

$<1/3$ Spherical micelles
$1/3 \sim 1/2$ Cylindrical
$1/2 \sim 1$ Lamellar or vesicles
>1 Inverse (reversed) micelles

The critical packing parameter is a useful parameter in aggregate design, as it can be modified for a given ionic surfactant by the addition of an electrolyte, addition of a co-surfactant, change in temperature, change in counter ion, or insertion of unsaturated or branched chains. Controlling aggregate architecture has enormous potential in many academic and industrial areas.

1.3.2 Surfactant-Stabilized Soft Dispersed Systems

Emulsification and foamation are the most versatile properties of surfactants for practical applications. Emulsions and foams are of fundamental importance in many applications and various fields of science and technology such as the food, pharma-

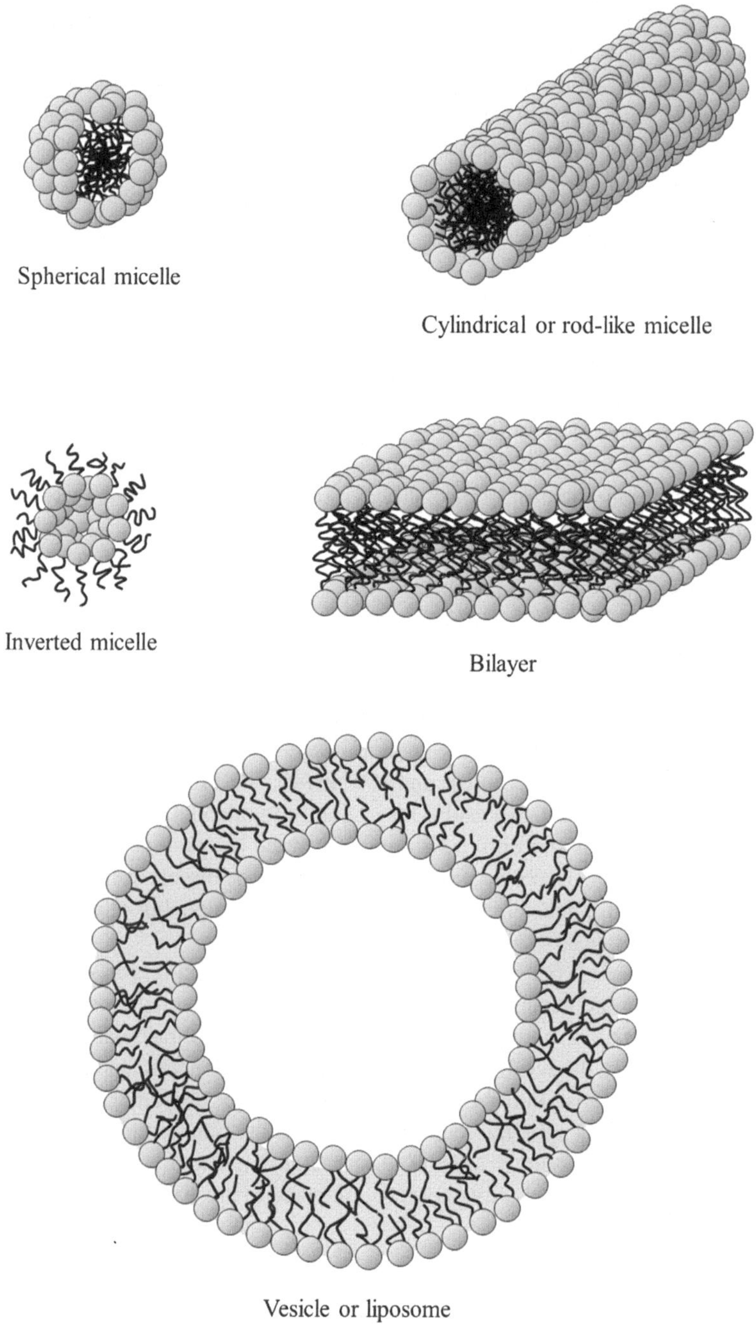

Fig. 1.13 Structure of surfactant aggregates

ceutical, and cosmetic industries. Two immiscible liquids or an immiscible liquid and a gas cannot form emulsions/foams, thus macrophase separation immediately occurs. In order to stabilize these dispersed systems, a surfactant is required. Surfactants can serve as these dispersion stabilizing agents (note that solid particles can also function as emulsifying agents as well as foaming agents). There are many great books and reviews devoted to emulsions and foams [44–47].

1.3.2.1 Emulsions

An emulsion is a stable suspension of liquid droplets within a continuous immiscible liquid. Emulsions can be classified into three types based on the size of the dispersed droplets, namely macroemulsions, miniemulsions, and microemulsions.

Macroemulsions are the most well-known milky emulsions, consisting of dispersed droplets with diameters of >400 nm [45]. Their opaque appearance is due to light scattering from the droplets, which have sizes on the order of the wavelength of visible light or larger. Macroemulsions are divided into two types based on the nature of the dispersed phase: oil-in-water (O/W) and water-in-oil (W/O) emulsions. Generally speaking, O/W emulsions are prepared using surfactants that are more soluble in water than in the oil phase, whereas W/O emulsions are produced using surfactants that are more soluble in the oil than in the water phase (Bancroft rule [48]). Macroemulsions are kinetically stabilized and they form only if energy is applied to the system. In other words, they are not thermodynamically stable.

Miniemulsions are the blue-white semiopaque emulsions consisting of dispersed droplets with diameters between 100 and 400 nm [49, 50]. Mixed emulsifier systems, comprising an ionic emulsifier and a long-chain fatty alcohol in concentrations of 1–3 wt% by weight based on the oil phase, are used to create stable oil-in-water miniemulsions. Water-insoluble molecules, which can be dissolved in the oil phase, are utilized as hydrophobes in order to prevent Ostwald ripening of oil droplets in oil-in-water emulsions.

Microemulsions are the emulsions consisting of dispersed droplets with diameters <100 nm that are generally obtained upon mixing the ingredients gently without vigorous agitation, high-pressure homogenization or ultrasonication [51–53]. Microemulsions are thermodynamically stable and their structure and properties do not depend on the preparation method or the time. Since the droplet diameter is below the visible light wavelength, microemulsions are usually transparent. In order to prepare microemulsions, a larger amount of surfactants (sometimes twice the weight based on the oil phase) is usually required, compared to those of macroemulsion and miniemulsion systems.

1.3.2.2 Foams

There are some similarities between aqueous foams (air-in-water dispersed systems) and macroemulsions, which is not surprising because a foam can be thought of as an emulsion in which the dispersed phase is a gas. Both systems consist of a dispersion of an immiscible state of matter in a continuous liquid phase. The oil–water and air–water interfaces stabilized with surfactant are soft and deformable. Both systems spontaneously macrophase separate unless there is an interfacial film that produces steric and/or electrical barriers against coalescence of the dispersed phase. In other words, neither system is thermodynamically stable, but they are kinetically trapped. The lifetime of liquid foams is largely determined by the repulsion between surfactants and the viscosity of the liquid. Their decay is driven by drainage caused by the negative Laplace pressure in the Plateau borders. There are also differences between foams and macroemulsions. The surfactants that stabilize foams cannot dissolve into the dispersed gas phase, while they can partially dissolve in the dispersed oil phase for O/W emulsions. In foams, only the liquid can act as the continuous phase, while both oil and water can work as the continuous phase in macroemulsions to form W/O and O/W emulsions, respectively. (Note that air-in-water foams and water-in-air dry water/liquid marbles can be prepared using particulate stabilizers [54, 55]).

References

1. Israelachvili J (1991) Intermolecular & surface forces. Academic Press, London
2. Wennerström E (1999) The colloidal domain. Wiley-VCH, New York
3. Earnshaw JC, Hugehs J (1992) Surface-induced phase transition in normal alkane fluid. Phys Rev A 46:R4494
4. Ocko BM, Wu XZ, Sirota EB, Sinha SK, Gang O, Deutsch M (1997) Surface freezing in chain molecules: normal alkanes. Phys Rev E 55:3164
5. Gang O, Wu XZ, Ocko BM, Sirota EB, Deutsch M (1998) Surface freezing in chain molecules. II. Neat and hydrated alcohols. Phys Rev E 58:6086
6. Hönig D, Möbius D (1991) Direct visualization of monolayers at the air-water interface. J Phys Chem 95:4590
7. Hénon S, Meunier J (1991) Microscope at the Brewster angle: direct observation of first-order phase transition in monolayers. Rev Sci Instrum 62:936
8. Mandelsohn R, Brauner JW, Gericke A (1995) External infrared reflection absorption spectrometry of monolayer films at the air-water interface. Ann Rev Phys Chem 46:305
9. Lei QL, Bain CD (2004) Surfactant-induced surface freezing at the alkane-water interface. Phys Rev Let 92:176103-1
10. Pershan PS (1990) Structure of surfaces and interfaces as studied using synchrotron radiation, Liquid surfaces. Faraday Disc Chem Soc 89:231
11. Als-Nielsen J, Jacquemain D, Kjaer K, Leveiller F, Lahav M, Leiserowitz L (1994) Principles and applications of grazing incidence X-ray and neutron scattering from ordered molecular monolayers at the air-water interface. Phys Rep 246:251
12. Kjaer K (1994) Some simple ideas on X-ray reflection and grazing-incidence diffraction from thin surfactant films. Phys B 198:100
13. Penfold J, Thomas RK (1990) The application of the specular reflection of neutrons to the study of surfaces and interfaces. J Phys Cond Matt 2:1369
14. Bowers J, Zarbakhsh A, Webster JRP, Hutchings LR, Richards RW (2001) Neutron reflectivity studies at liquid-liquid interfaces: methodology and analysis. Langmuir 17:140
15. Vollhardt D, Fainerman VB, Emrich G (2001) Dynamics and phase transition in adsorbed monolayers of sodium dodecyl sulfate/dodecanol mixtures. J Phys Chem B 104:8536

16. Lee EM, Thomas RK, Penfold J, Ward RC (1989) Structure of aqueous decyltrimethylammonium bromide solutions at the air water interface studied by the specular reflection of neutrons. J Phys Chem 93:381
17. Takiue T, Kawagoe S, Muroi S, Murakami R, Ikeda N, Aratono M, Tanida H, Sakane M, Harada M, Watanabe I (2003) Surface density measurement of the bromide ion by the total-reflection X-ray absorption fine structure technique at the air/aqueous dodecyltrimethylammonium bromide solution interface. Langmuir 19:10803
18. Barthlott W, Neinhuis C (1997) Purity of the sacred lotus, or escape from contamination in biological surfaces. Planta 202:1–8
19. Feng L, Zhang Y, Xi J, Zhu Y, Wang N, Xia F, Jiang L (2008) Petal effect: a superhydrophobic state with high adhesive force. Langmuir 24:4114–4119
20. Gao X, Yan X, Yao X, Xu L, Zhang K, Zhang J, Yang B, Jiang L (2007) The dry-style antifogging properties of mosquito compound eyes and artificial analogues prepared by soft lithography. Adv Mater 19:2213–2217
21. Gómez-Suárez C, Bruinsma GM, Rakhorst G, van der Mei HC, Busscher HJ (2002) Hydrophobicity of peritoneal tissues in the rat. J Colloid Interface Sci 253:470–471
22. Gao X, Jiang L (2004) Water-repellent legs of water striders. Nature 432:36
23. Young T (1805) An essay on the cohesion of fluids. Phil Trans Roy Soc Lond 95:65–87
24. Wenzel RN (1936) Resistance of solid surfaces to wetting by water. Ind Eng Chem 28:988–994
25. Cassie ABD, Baxter S (1944) Wettability of porous surfaces. Trans Faraday Soc 40:546–551
26. Lafuma A, Quéré D (2003) Superhydrophobic states. Nat Mat 2:457–460
27. Ishino C, Okumura K, Quéré D (2004) Wetting transitions on rough surfaces. Europhys Lett 68:419–425
28. Nonomura Y, Morita Y, Hikima T, Seino E, Chida S, Mayama H (2010) Spreading behavior of water droplets on fractal agar gel surfaces. Langmuir 26:16150–16154
29. Nonomura Y, Chida S, Seino E, Mayama H (2012) Anomalous spreading with Marangoni flow on agar gel surfaces. Langmuir 28:3799–3806
30. Onda T, Shibuichi S, Satoh N, Tsujii K (1996) Super-water-repellent fractal surfaces. Langmuir 12:2125–2127
31. Hosono E, Fujihara S, Honma I, Zhou H (2005) Superhydrophobic perpendicular nanopin film by the bottom-up process. J Am Chem Soc 127:13458–13459
32. Lau KKS, Bico J, Teo KBK, Chhowalla M, Amaratunga GAJ, Milne WI, McKinley GH, Gleason KK (2003) Superhydrophobic carbon nanotube forests. Nano Lett 3:1701–1705
33. Yabu H, Takebayashi M, Tanaka M, Shimomura M (2005) Superhydrophobic and lipophobic properties of self-organized honeycomb and pincushion structures. Langmuir 21:3235–3237
34. Tsujii K, Yamamoto T, Onda T, Shibuichi S (1997) Super oil-repellent surfaces. Angew Chem Int Ed 36:1011–1012
35. Tuteja A, Choi W, Ma M, Mabry JM, Mazzella SA, Rutledge GC, McKinley GH, Cohen RE (2007) Designing superoleophobic surfaces. Science 318:1618–1622
36. Wang R, Hashimoto K, Fujishima A, Chikuni M, Kojima E, Kitamura A, Shimohigoshi M, Watanabe T (1997) Light-induced amphiphilic surfaces. Nature 388:431–432
37. Butt H-J, Graf K, Kappl M (2006) Physics and chemistry of interfaces, 2nd Revised and enlarged edition. Wiley-VCH, Berlin
38. Rosen MJ (2004) Surfactants and interfacial phenomena, 3rd edn. Wiley-Interscience, New Jersey
39. Israelachvili JN (2011) Intermolecular and surface forces, 3rd edn. Academic Press, USA
40. Schick MJ, Fowkes FM (Consulting ed) (1967–2011) Surfactant science series, vol 1–152
41. Rosen MJ (ed) (1984) Structure/performance relationships in surfactants. ACS Symposium Series, vol 253
42. Liu S, Armes SP (2001) Recent advances in the synthesis of polymeric surfactants. Curr Opin Colloid Interface Sci 6:249–256
43. Evans DF (1988) Self-organization of amphiphiles. Langmuir 4:3–12
44. Colans C, Kunieda H (1996) Industrial applications of microemulsions. Marcel Dekker, New York

45. Becher P (1965) Emulsions: theory and practice. Reinhold Publishing, New York
46. Pugh RJ (1996) Foaming, foam films, antifoaming and defoaming. Adv Colloid Interface Sci 64:67–142
47. Isenberg C (1992) The science of soap films and soap bubbles. Dover Publications, New York
48. Bancroft WD (1913) The theory of emulsification, V. J Phys Chem 17:501–519
49. Ugelstad JM, El-Asser S, Vanderhoff JW (1976) Emulsion polymerization: initiation of polymerization in monomer droplets. J Polym Sci Polym Lett 11:503–513
50. Sagitani H (1981) Making homogeneous and fine droplet O/W emulsions using nonionic surfactants. J Am Oil Chem Soc 58:738–743
51. Hoar TP, Schulman JH (1943) Transparent water-in-oil dispersions: the oleopathic hydromicelle. Nature 152:102–103
52. Winsor PA (1948) Hydrotropy, solubilization, and related emulsification processes. Trans Faraday Soc 44:376–398
53. Sjöblom J, Lindberg R, Friberg SE (1996) Microemulsions—Phase equilibria characterization, structures, applications and chemical reactions. Adv Colloid Interface Sci 95:125–287
54. Fujii S, Murakami R (2008) Microparticles as foam and liquid marble stabilizers. KONA Powder Particle J 26:153–166
55. Fujii S, Nakamura Y (2017) Stimuli-responsive bubbles and foams stabilized with solid particles. Langmuir 33:7365–7379

Part II
Design of Soft Interface (Synthesis and Processing)

Chapter 2
Molecular Design of Soft Interface

Shin-ichi Yusa and Syuji Fujii

2.1 Polymerization Method

Polymer synthesis techniques are very important to create and design soft interfaces. In particular, the strict control of polymer architecture is an important issue. Homopolymers, random copolymers, and block copolymers with well-defined molecular weight, structure, and narrow molecular weight distribution can be prepared via living or controlled/living polymerization techniques. Free radical polymerizations are preferable to ionic polymerizations on economic reasons. There are various limitations of monomers and solvents for ionic living polymerization methods including anionic and cationic polymerization in comparison with controlled/living radical polymerization (CLRP) methods. The CLRP techniques are relatively easy to perform, because it does not require any strict purification of the reagents and is tolerant to many functional groups.

These techniques can be utilized to prepare various kinds of polymer architectures including statistical copolymer, block copolymer, gradient copolymer, graft copolymer, cross-linked polymer, hyperbranched polymer, and star-shaped polymer. The key feature of CLRP is the existence of an equilibrium state between active and dormant species, which allows a controlled growth of chains while retaining a sufficiently low concentration of propagating radicals to minimize termination and side reactions (Fig. 2.1). The CLRP methods that have widely used include iniferter, nitroxide-mediated radical polymerization (NMP), atom transfer radical polymerization (ATRP), reversible addition-fragmentation chain transfer (RAFT)

S. Yusa (✉)
Department of Applied Chemistry, Graduate School of Engineering, University of Hyogo, 2167 Shosha, Himeji, Hyogo 671-2280, Japan
e-mail: yusa@eng.u-hyogo.ac.jp

S. Fujii
Department of Applied Chemistry, Faculty of Engineering, Osaka Institute of Technology, 5-16-1 Omiya, Asahi-Ku, Osaka 535-8585, Japan

M. Maeda et al. (eds.), *Molecular Soft-Interface Science*,
https://doi.org/10.1007/978-4-431-56877-3_2

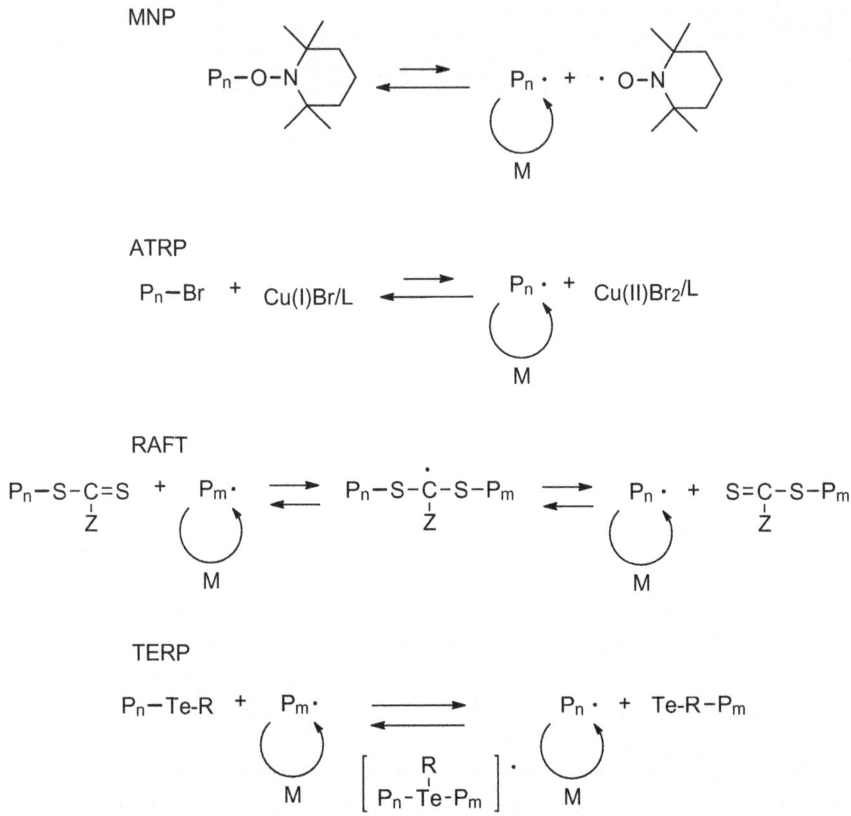

Fig. 2.1 Mechanisms for controlled/living radical polymerizations for nitroxide mediated radical polymerization (NMP), atom transfer radical polymerization (ATRP), reversible addition-fragmentation chain transfer (RAFT), and organotellulium-mediated radical polymerization (TERP) methods

radical polymerization, and organotellurium-mediated living radical polymerization (TERP).

Iniferter is shorthand for initiator-transfer agent-terminator. Iniferter is used to describe a group of compounds which under thermal or photoinitiation polymerization systems. The propagating radicals are converted reversibly into dormant chains and eventually participate in the termination of the propagating radical chains.

The early method of NMP is started by conventional free radical initiators in the presence of stable nitroxide radicals such as 2,2,6,6-tetramethylpiperidinyl-1-oxy (TEMPO). Subsequent polymerizations involve reversible attachment of the mediating compounds to the propagating chains.

The general mechanism of ATRP is that organic halides initiate the polymerization of monomers. During the polymerization, the propagating radicals react with the metal halide, which should be a deactivator of radical polymerization to reform the

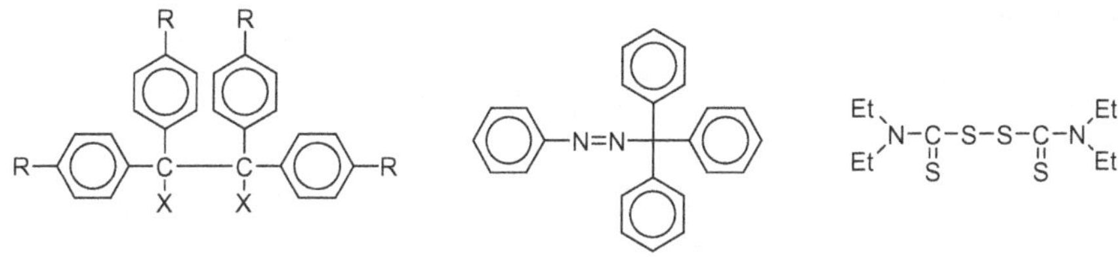

Fig. 2.2 Examples of thermal- and photoiniferters

lower oxidation state metal complex and a polymer chain with a halogen end group. ATRP is the most rapidly developing areas in polymer synthesis.

RAFT is one of the most extensively studied CLRP methods. RAFT polymerization based on the principle of degenerative chain transfer. The process involves the conventional free radical polymerization in the presence of a chain transfer agent.

The mechanism of TERP contains reversible termination and degenerative chain transfer. TERP can control the polymerization of conjugate monomers and unconjugate monomers using the same TERP agent.

2.2 Iniferter

Iniferter is used to describe a group of compounds which under thermal and photoinitiation of radical polymerization of vinyl monomers, convert reversibly the propagating radicals into dormant chains and eventually participate in the termination of the propagating radicals (Fig. 2.2) [1–3].

Conventional free radical polymerization is terminated by bimolecular reactions between propagating radicals including recombination and disproportionation. Recombination and disproportionation reactions are determined by a kind of monomers, solvent, and reaction conditions. For example, radical polymerization of styrene is almost terminated by the recombination process. On the other hand, in the case of radical polymerization of methyl methacrylate, disproportionation is dominant at high-temperature due to the α-methyl group. When initiator which tends to occur chain transfer to itself and termination of primary radicals is used for radical polymerization, it may be formed polymer chains have initiator fragments at both the chain ends.

First, methyl methacrylate is polymerized by thermal iniferters (Fig. 2.2a–c). The photoinduced polymerization of various vinyl monomers are performed by photoiniferters (Fig. 2.2d). Moreover, iniferters have been used for the preparation of block copolymers, since reversible transformations of propagating chains into dormant chains enable the formation of sequences of monomer units during their copolymerization [4].

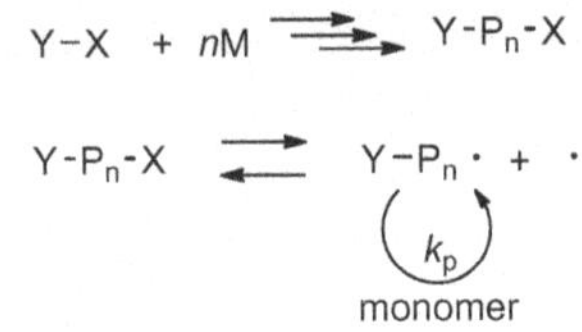

Fig. 2.3 Mechanism of iniferter

If monomers are inserted into the iniferter bond (Y–X), the polymerization proceeds in accordance with the living mechanism. Figure 2.3 shows the mechanism of iniferter radical polymerization. Y–X is iniferter (sometimes Y = X), M is monomer, Y–P_n–X is dormant chain, and Y–P_n• is propagating radical. In principle the leaving radical group, X• don't have initiating ability, which can only react with propagating radicals, Y–P_n•. The simultaneous formation of a fraction of homopolymers cannot be avoided, because the iniferter process contains termination reactions. The molecular weight of polymer formed from iniferter process increases with increasing polymerization time. Sometimes the conversion is low. Polymerization initiated via iniferter is often suite slow and may yield polymers with broad molecular weight distributions.

All of photoiniferters were based on the readily dissociating *N,N'*-diethyldithiocarbamate group [5]. Apparently, the photoiniferters may be stabilized by its removal of through a photoinduced transfer to thiol. In spite of the drawbacks of the iniferters method, as compare to other controlled/living radical polymerization methods, their use as initiators of radical polymerizations may sometimes be of interest for the introduction of functional end-groups and for the controlled surface grafting of oligomers, polymers, and copolymers [6].

2.3 Stable Free Radical Polymerization (NMP)

Nitroxide radical compounds are stable which are able to trap carbon-centered radicals at a nearly diffusion-controlled rate. Generally, at low temperature, the formed alkoxyamine is stable and the trapping reaction corresponds to an irreversible termination reaction [7–9]. However, at high temperature, the covalent bond between carbon and oxygen undergoes homolytic cleavage, leading back to the propagating radical and to the nitroxide. In NMP, dormant polymeric alkoxyamines dissociate to produce a propagating radical and a persistent nitroxide radical. The former adds to monomer and re-couples with persistent radical. The equilibrium between propagating radical and dormant alkoxyamine is the key process in NMP. It is important that the stable nitroxide radicals are capable of the reversible termination reactions, but do not initiate polymerizations.

In the propagating radical (P_n•), it is added by the monomer, wherein the dormant state (P_n–X), the reactive chain end is blocked with no polymerization allowed (Fig. 2.4). Here, X• indicates the stable nitroxide radicals, k_p is the propagation rate

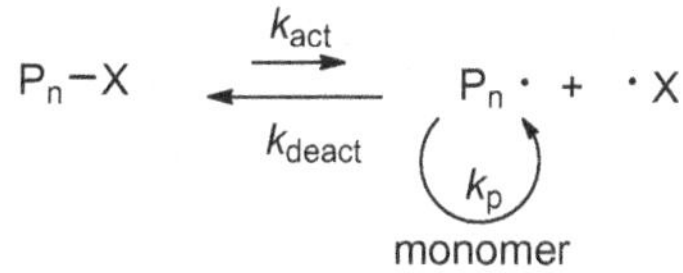

Fig. 2.4 NMP equilibrium

constant, k_{act} is the activation constant, and k_{deact} is the deactivation constant. A number of activation and deactivation cycles and a low P_n• allow all the chains to grow slowly and simultaneously, suppressing irreversible bimolecular termination. The propagation rate (R_p) of NMP is independent of nitroxide radical [10]. The consumption of [X•] and [P_n•] can be expressed following equations:

$$\mathrm{d}[X\bullet]/\mathrm{d}t = k_{act}[\mathrm{P_n} - \mathrm{X}] - k_{deact}[\mathrm{P_n}\bullet][\mathrm{X}\bullet] \quad (2.1)$$

$$\mathrm{d}[\mathrm{P_n}\bullet]/\mathrm{d}t = k_{act}[\mathrm{P_n} - \mathrm{X}] - k_{deact}[\mathrm{P_n}\bullet][\mathrm{X}\bullet] + R_i - k_t[\mathrm{P_n}\bullet]^2 \quad (2.2)$$

where t is the reaction time, R_i is the rate of initiation, k_t is the rate constant of termination. If R_i is nonzero, the system should reach the stationary state in which d[P_n•]/dt = d[X•]/dt = 0.

$$[\mathrm{P_n}\bullet] = (R_i/k_t)^{1/2} \quad (2.3)$$

$$[\mathrm{X}\bullet] = (k_{act}/k_{deact})[\mathrm{P_n} - \mathrm{X}]/[\mathrm{P_n}\bullet] \quad (2.4)$$

Therefore, the stationary concentration of P_n• is determined by only the balance of the initiation and termination rate constant from Eq. 2.4. X• depends on not only on this stationary value of [P_n•] but also on [P_n–X] and activation and deactivation rate constants from Eq. 2.4. Therefore, R_p can be expressed as the following equation.

$$R_p = k_p[\mathrm{P_n}\bullet][\mathrm{M}] = k_p(R_i/k_t)^{1/2}[\mathrm{M}] \quad (2.5)$$

Equation 2.5 indicates R_p is independent of [P_n–X]. The kinetic of NMP using styrene and TEMPO was studied [11]. The pseudo-first-order kinetic plot for the styrene polymerization with and without PS-TEMPO adduct shows that the R_p value of the nitroxide system is equal to that of nitroxide-free thermal system at low conversion (<30%).

The existence of the reversible dissociation of the dormant will be evidenced by showing the existence of the equilibrium constant K.

$$K = k_{act}/k_{deact} = [\mathrm{P_n}\bullet][\mathrm{X}\bullet]/[\mathrm{P_n} - \mathrm{X}] \quad (2.6)$$

In situ electron spin resonance (ESR) measurements for NMP of styrene in the presence of TEMPO indicates that [P_n•] is on the order of 10^{-8} mol L^{-1} and [X•]

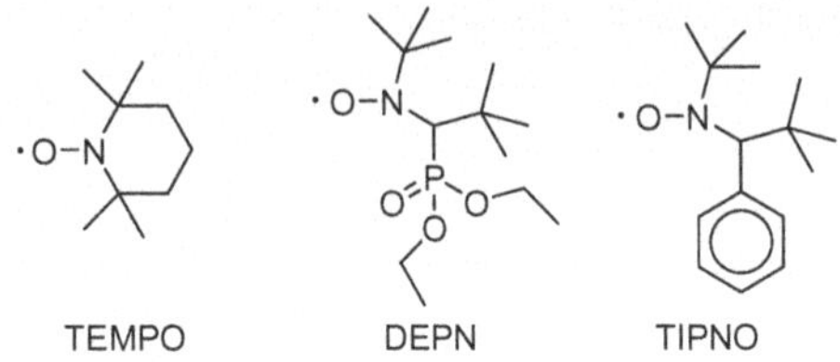

Fig. 2.5 Chemical structures of stable nitroxide radical compounds

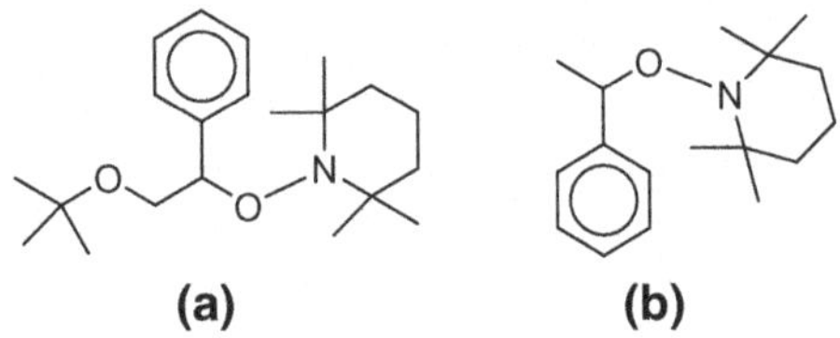

Fig. 2.6 Chemical structures of universal initiators for NMP

is on the order of 10^{-5} mol L^{-1} [11]. The initial concentration of $[P_n\text{-}X]$ is 3.6×10^{-2} mol L^{-1}, which is much larger than both $[P_n\bullet]$ and $[X\bullet]$. The K value can be estimated to be 2.1×10^{-11} mol L^{-1}, independent of polymerization time.

The main nitroxides (Fig. 2.5) are TEMPO [12, 13], *N-tert*-butyl-*N*-(1-diethylphosphono-(2,2-dimethylpropyl)) nitroxide (DEPN) [14], and *N-tert*-butyl-*N*-(1-phenyl-2-(methylpropyl))nitroxide (TIPNO) [9]. TEMPO is successfully applied for styrene and its derivatives. DEPN and TIPNO allow the range of monomer including styrene and acrylate derivatives to be very significantly expanded and the polymerization time to be reduced. The initiation process is performed in two different methods by using either bicomponent system with an ordinary radical initiator such as benzoyl peroxide (BPO) with the free nitroxide, or a monocomponent system based on a preformed alkoxyamine. Now the later is the most popular method because it allows the kinetics and molar mass to be tuned precisely (Fig. 2.6).

Universal initiators can be used for NMP. The reversible hemolytic dissociation of Fig. 2.6a (2-*tetrt*-butoxy-1-phenyl-1-(1-oxy-2,2,6,6-tetramethyl piperidinyl)ethane) to yield the corresponding carbon-centered radical and nitroxide (TEMPO). This fundamental step is added to a conventional free radical polymerization. The universal initiator (Fig. 2.6b) is prepared by hydrogen atom abstraction from ethylbenzene and trapping the resultant radical with TEMPO [15]. Thermal decomposition of II shows a continuous formation of TEMPO by ESR measurements. Decomposition products revealed the formation of styrene monomer. Figure 2.6b decomposes in the temperature range utilized for NMP of styrene at a rate comparable to the styrene conversion rate. End group purity achieved by NMP of styrene decreased with increasing both molecular weight and conversion. The termination process would limit the ability of NMP for the preparation of high molecular weight (>100,000) polystyrene with narrow molecular weight distributions.

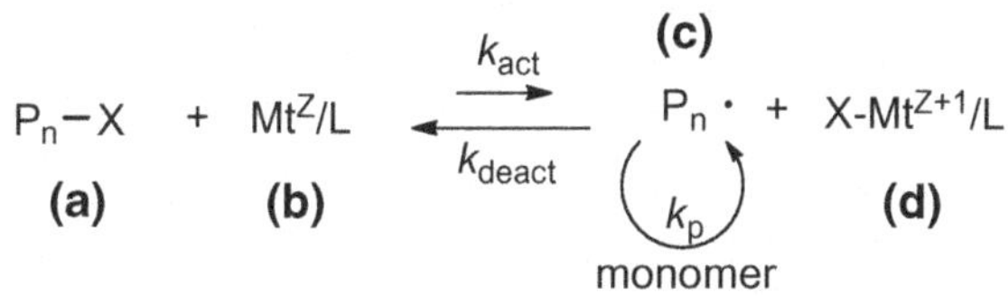

Fig. 2.7 The ATRP equilibrium state, which resides predominantly to the dormant side (left), includes dormant P_n–X (Fig. 2.7a), catalyst composed of a transition metal with ligand (Mt^Z/L, Fig. 2.7b), active radical ($P_n\bullet$, Fig. 2.7c), and oxidized catalyst (Fig. 2.7d). The reaction rates activation, deactivation, and propagation are labeled with k_{act}, k_{deact}, and k_p, respectively

2.4 Atom Transfer Radical Polymerization (ATRP)

ATRP is based on the transfer of a halogen atom from a dormant initiator of polymeric chain to a transition metal salt. A ligated salt has catalyzed the transfer. The transition metal is oxidized when the halogen atom is transferred and a propagating radical is generated.

Figure 2.7 shows the equilibrium state of ATRP. The mechanism consists of phenomenologically related initiation and propagation processes. These sequences are comprised of an atom transfer equilibrium along with radical addition to the monomer. Termination reaction by radical coupling and disproportionation are included in the mechanism because of the magnitude of the associated rate constant, however, only a few percents of the polymer chains undergo termination. The equilibrium reaction of ATRP is the reversible transfer of halide (X) from a dormant species P_nX (dormant polymer chain) to transitional metal complex Mt^Z/L (L is a ligand). The dormant species (Fig. 2.7a) periodically react with the rate constant of activation (k_{act}) with transition metal complex with lower oxidation state. Mt^Z represents the transition metal species in oxidation state Z. The deactivator (Fig. 2.7d) reacts with the propagating radical in a reverse reaction (k_{deact}) to reform the dormant species and the activator (Fig. 2.7b). Polymerization is propagated by the addition of monomer to thus generated propagating radicals. The catalyst is regenerated by reduction of the oxidized transition metal complex, while the propagating radical chain is converted into a dormant chain. ATRP can be applied to a broad range of vinyl monomers.

ATRP is catalytic process and can be mediated by many redox-active transition metals such as Cu^I [16, 17], Ni^{II} [18, 19], Ru^{II} [20], Fe^{II} [21]. The reaction rate of ATRP derived from Fig. 2.7 by neglecting the contribution of termination, assuming that initiation is complete and using the fast equilibrium approximation. When Cu(I) is used as catalysis of the redox-active transition metal, the rate of propagation can be expressed as follow:

$$R_p = k_p[M][P_n\bullet] = k_p[M]K_{ATRP}\left(\frac{[P_nX][Cu^I/L]}{[XCu^{II}/L]}\right) \tag{2.7}$$

$$K_{\mathrm{ATRP}} = \frac{k_{\mathrm{act}}}{k_{\mathrm{deact}}} \tag{2.8}$$

where $[P_nX]$ is the concentration of dormant chains which corresponds to the initial concentration of the initiator ($[RX]_0$). The structure of the ligand, monomer, and dormant species as well as reaction conditions such as solvent, temperature, and pressure strongly influence the values of the rate constants, k_{act} and k_{deact} [22]. The R_p value increases with increasing K_{ATRP}, however, under some conditions, the polymerization cannot be controlled due to radical termination.

ATRP and in all other CLRP processes, termination reactions are not entirely eliminated. Therefore, chains will continuously terminate resulting in the increase the persistent radical concentration and the progressive decrease of the propagating radical concentration which may lead to some deviation from the first-order kinetics. However, the contribution of the termination process under appropriate conditions is small and the concentration of radicals is reduced less than 10% during consumption of nearly all monomer and, therefore, deviation from the first-order kinetic may not be detected.

The propagating radical concentration decreased to 10^{-7} M and the deactivator concentration increases approximately 10^{-3} M. Under these conditions, the termination rate ($R_t = k_t[P_n\bullet]^2$) will be significantly slower than the rate at which propagating radicals will react with the Cu^{II}/L ($k_{\mathrm{deact}}[P_n\bullet][Cu^{II}/L]$) in a deactivation process, and a CLRP will ensure. Less than 10% of the polymer chains are terminated during the initial short nonstationary process, but the majority of the chains (>90%) continue the polymerization successfully [23].

In a polymerization based on the ATRP catalytic cycle, the control of the polymerization and of the resulting polymers will depend on the stationary concentration of the propagating radicals and the relative rates of propagation and deactivation. During one activation step, any number of monomer units can be added to the polymer chain with varying effects upon the polydispersity index (PDI) of the polymers formed.

$$\mathrm{PDI} = 1 + \frac{1}{\mathrm{DP}} + \left(\frac{k_p[P_nX]}{k_{\mathrm{deact}}[XCu^{II}/L]}\right)\left(\frac{2}{\rho} - 1\right) \tag{2.9}$$

The PDI value is affected by the concentrations of dormant species ($[P_nX]$) and deactivator ($[XCu^{II}/L]$), the rate constant of propagation (k_p) and deactivation (k_{deact}), and the monomer conversion (ρ) [24]. The PDI values are higher for shorter chains relative to longer chains due to the growth of smaller chains involves fewer activation–deactivation steps and therefore fewer opportunities for exchange and controlled growth. The final PDI values should be higher for higher values of ratio (k_p/k_{deact}). Therefore, under similar conditions, the polymerization of acrylates yields higher PDI than that of styrene, because k_p for acrylates is much larger than that of styrene.

The main role of alkyl halide (RX) species is to dictate the number of initiated chains. The polymerization rates in ATRP are first-order with respect to the concentration of RX, and the molecular weights reciprocally with the initial concentration

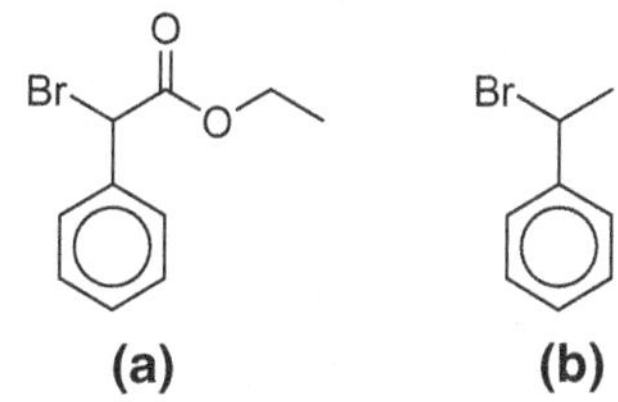

Fig. 2.8 Chemical structures of initiators for ATRP

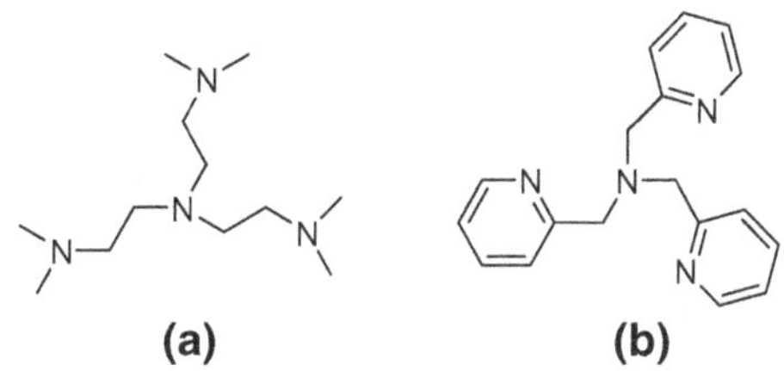

Fig. 2.9 Chemical structures of ligands for ATRP

of RX (Fig. 2.8). Alkyl halide reactivates follow the order I > Br > Cl. The reactivity of RX follows the order of 3° > 2° > 1°, in agreement with bond dissociation energy needed for hemolytic bond cleavage. Stability of radicals is enhanced by the presence of a cyano, phenyl, or ester group. The most active initiator is ethyl α-bromophenylacetate (Fig. 2.8a), with combined activation effect of both phenyl and ester groups. Ethyl α-bromophenylacetate is more than 10,000 times more active than 1-phenylethyl bromide (Fig. 2.8b) [25].

The catalyst complex Fig. 2.7b pulls out the halogen from the initiator to active the polymerization. Therefore, the catalyst has an important influence on the reaction rates of the equilibrium. The ligand select will have a profound effect on k_{act} and k_{deact}, which will cause a difference in the rate of polymerization. Tris(2-dimethylaminoethyl)amine (Fig. 2.9a, Me_6TREN) and tris(2-pyridylmethyl)amine (Fig. 2.9b, TPMA) are most active ligands for Cu complex (Fig. 2.9) [26].

Polymerization of acidic monomers such as acrylic acid and methacrylic acid cannot control via ATRP because acidic monomers protonate ligands to be unstable of the catalyst complex. Dienes cannot be polymerized via ATRP because dienes displace ligands and generate less redox active complexes to destroy the catalyst complexes.

2.5 Reversible Addition-Fragmentation Chain Transfer (RAFT) Radical Polymerization

Chain transfer agent (CTA) used in RAFT process has a dithioester (S = C–S) group and various kinds of R and Z groups (Fig. 2.10). The nature of R and Z groups is important for control of polymerization via RAFT. CTAs include dithioesters (Z = alkyl or phenyl), trithiocarbonates (Z = SR'), xanthates (Z = OR'), and dithiocar-

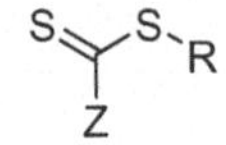

Fig. 2.10 A typical example of chemical structure of chain transfer agent

bamates (Z = NR'R"). Selection of CTA for the monomers and reaction conditions is crucial for the success of a RAFT radical polymerization. Polymerization of most monomers can be well controlled to provide minimal retardation and high fraction of living chains by using appropriate CTA. Generally, dithioesters and trithiocarbonates are suitable for polymerization of conjugate monomer which is more activated monomers (MAM) such as methyl methacrylate, methyl acrylate, styrene, and acrylonitrile. On the other hand, xanthates and dithiocarbamates are suitable for controlled polymerization of unconjugated monomers which is less activated monomers (LAM) such as vinyl acetate, *N*-vinylpyrrolidone, and *N*-vinylcarbazole.

The mechanism of RAFT radical polymerization comprises the addition–fragmentation equilibrium shown in Fig. 2.11 [27]. In the RAFT mechanism, radicals are neither formed nor destroyed. Thus, RAFT polymerization process will not take place without an external supply of radicals from a radical initiator. The RAFT equilibrium state has no direct influence on the rate of polymerization beyond that caused by the reduction in molecular weight and molecular weight distribution. It should be noted that the termination process between two propagating radicals is not directly suppressed by the RAFT process. The detailed kinetics of actual RAFT is complicated, because there maybe side reactions in each process [28–30]. If fragmentation is slow, the intermediate radicals (Fig. 2.11b or e) is consumed in side reactions, or reinitiation is slow or inefficient, and then retardation of inhibition occur. Optimal control of RAFT process requires selection of an appropriate CTA (Fig. 2.11a) for monomers to be polymerized. The Z and R groups both play important roles in determining the outcome of RAFT polymerization. By determining the rate of addition (k_{add}) and fragmentation (k_{β}), they control the efficiency of chain transfer and retardation.

In conventional free radical polymerization, all propagating radicals (P_m• and P_n•) are generated from an initiator. On the other hand, in the RAFT process the most propagating radicals are generated from the leaving group of R• (Fig. 2.11d). In initialization process, propagating radicals are added to CTA to generate intermediate radicals (Fig. 2.11b). Reversible β-fragmentation of Fig. 2.11b generate R•, which is a reversible chain transfer reaction from P_n• to R•. The properties of CTA (Fig. 2.11a) can be defined in terms of two transfer coefficients, C_{tr} ($=k_{tr}/k_p$) and C_{-tr} ($=k_{-tr}/k_{iR}$) where the rate constants k_{tr} and k_{-tr} are defined in terms of the rate constants for radical addition, k_{add} and $k_{-\beta}$, and a partition coefficient (ϕ) as expressed in following equations.

$$k_{tr} = k_{add} \times \varphi = \frac{k_{add} \times k_{\beta}}{k_{-add} + k_{\beta}} \tag{2.10}$$

$$k_{-tr} = k_{-\beta}(1 - \varphi) = \frac{k_{-\,add} \times k_{-\beta}}{k_{-add} + k_{\beta}} \tag{2.11}$$

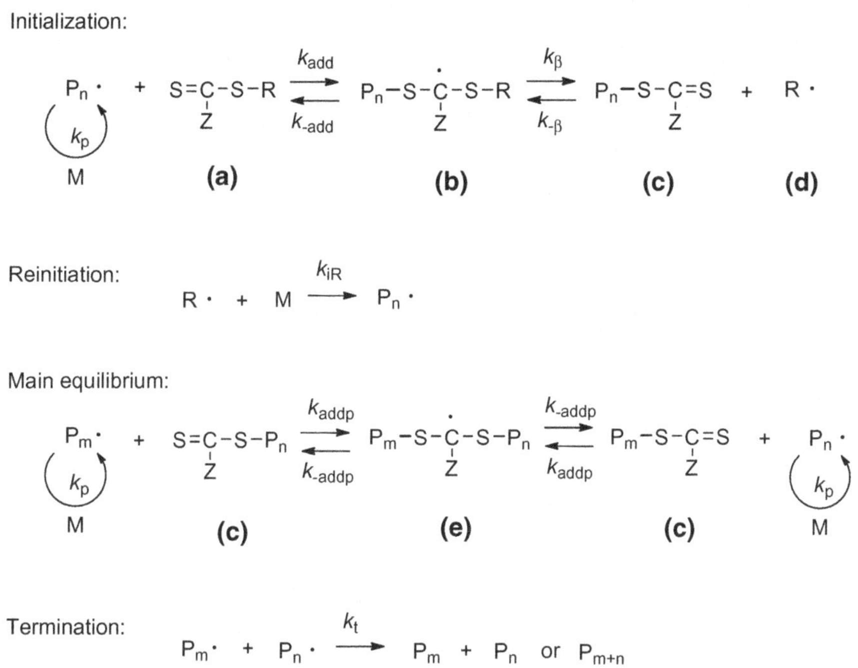

Fig. 2.11 Mechanism of reversible addition-fragmentation chain transfer (RAFT) radical polymerization

$$\varphi = \frac{k_\beta}{k_{-\text{add}} + k_\beta} \tag{2.12}$$

The partition coefficient ϕ indicates the preference for the intermediate radials Fig. 2.11b (or e) to fragment to products or return to starting materials. For effective CTA, R should be a good hemolytic leaving group with respect to the propagating radical. Therefore ϕ should be more than 0.5 for efficient control of RAFT polymerization. For macro-CTA (Fig. 2.11c) formed in RAFT homopolymerization, where m and n > 2, then $C_{tr} = C_{\text{-tr}}$ and ϕ will be 0.5.

$C_{tr}(P_nCTA)$ in main equilibrium state differ from $C_{tr}(CTA)$ in initialization state. In initialization sate the chain transfer occurs from $P_n\bullet$ to CTA, however, in main equilibrium sate propagating radicals ($P_m\bullet$ and $P_n\bullet$) cannot be distinguished. In the main equilibrium state, as the rate constants for addition and fragmentation are k_{addp} and $k_{\text{-addp}}$ ($=k_{addp}$), $C_{tr}(P_nCTA)$ can be expressed as follow:

$$C_{tr}(P_nCTA) = \frac{k_{\text{addp}} \times k_{-\text{addp}}}{k_p \times (k_{\text{addp}} + k_{-\text{addp}})} = \frac{k_{\text{addp}}}{2k_p} \tag{2.13}$$

Table 2.1 Apparent chain transfer constant (C_{trapp}) for RAFT polymerization of styrene

Z group	R group	Temperature (°C)	C_{trapp}	Reference
Ph	$C(CH_3)_2Ph$	60	>500	[31]
Ph	$C(CH_3)_2Ph$	60	25000	[33]
Ph	CH_2Ph	60	50	[31]
Ph	CH_2Ph	60–90	190	[34]
Ph	$C(CH_3)_2Ph$	110	29	[35]
CH_3	CH_2Ph	110	10	[35]
SC_4H_9	$CH(CH_3)Ph$	110	>100	[36]
OEt	$CH(CH_3)Ph$	60	0.65	[37]
OEt	$CH(CH_3)CO_2Et$	60	0.67	[37]
OEt	CH_2Ph	110	0.11	[35]
NEt_2	CH_2Ph	110	0.009	[38]
SCH_2Ph	CH_2Ph	80	53	[39]

To simplify $C_{tr}(P_nCTA)$ and $C_{tr}(CTA)$ are assumed to equal to apparent chain transfer constant (C_{trapp}). The degree of polymerization (DP) and molecular weight distribution (PDI) can be expressed as follows [32]:

$$\mathrm{DP} = \frac{[\mathrm{M}]_0}{[\mathrm{CTA}]_0} \times \frac{\rho}{1-(1-\rho)^{C_{trapp}}} \tag{2.14}$$

$$\mathrm{PDI} = 1 + \frac{1}{C_{trapp}} \tag{2.15}$$

where $[M]_0$ and $[CTA]_0$ are initial concentrations of monomer and CTA, and, ρ is conversion. Table 2.1 summarized C_{trapp} for RAFT radical polymerization of styrene. Polydispersity index (PDI) of the resulting polymer begins to be narrow in RAFT process at $C_{trapp} > 2$. When C_{trapp} is more than 10, the consumption rate of CTA is faster than that of monomer. The polymerization can be controlled because the concentration of propagating radicals decreases. PDI will be sufficiently narrow and degree of polymerization (DP) increases with increasing the conversion which can be observed. These observations indicate that the polymerization proceeds in accordance with the living mechanism. The predicted dependence of the DP and PDI of the polymer formed on monomer conversion and transfer coefficient for ideal polymerization with reversible chain transfer is calculated by simulation [31].

When CTA has phenyl as a Z group, isobutyronitrile or cumyl groups containing CTAs are a good candidate to polymerize MAM such as styrene and methyl methacrylate monomers [40, 41]. RAFT polymerization of LAM can be controlled using xanthate and dithiocarbamate CTA. Propagating radicals with a terminal MAM unit are less reactive in radical addition, and one of the more active CTA is required for good control of the polymerization. The poly(MAM) propagating radicals are rel-

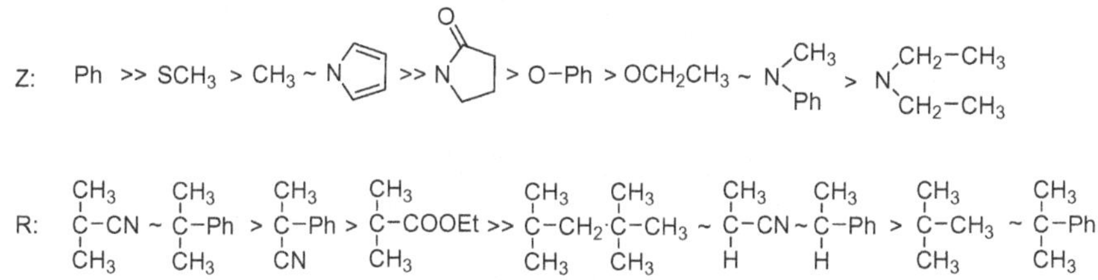

Fig. 2.12 Chemical structures of chain transfer agents. Z groups; the chain transfer constants decrease from left to right. R groups; the fragmentation reaction rates decrease from left to right

atively good hemolytic leaving groups. The more active CTAs such as the dithioesters and trithiocarbonates allow the preparation low polydispersity. On the other hand, xanthates and dithiocarbamates have lower transfer constants and provide poor control of the polymerization. Propagating radicals with a terminal LAM unit are highly reactive in radical addition. Addition to less active CTA such as dithiocarbamates and xanthates, it is sufficient that these CTAs have high transfer constants in LAM polymerization. However, poly(LAM) propagating radicals are relatively poor hemolytic leaving groups. When more active CTAs, such as dithioesters, are used in LAM polymerization, fragmentation is slow and inhibition or retardation is likely. General guidelines for selection of Z and R are shown in Fig. 2.12.

The R group of CTA must be a good hemolytic leaving group with respect to P_n•, such that the intermediate radical Fig. 2.11b, both fragments rapidly and partitions in favor of Fig. 2.11c and R•. The Ro must also be able to reinitiate polymerization efficiently. The chain transfer constant increases in the series primary < secondary < tertiary and with the introduction of substituents which are capable of delocalizing the radical center are consistent with the view. In the synthesis of block copolymers by RAFT polymerization comprising blocks of 1,1-disubstituted monomer and a monosubstituted monomer, 1,1-disubstituted monomer should be polymerized first [31].

There are some drawbacks for RAFT. The preparation of the corresponding CTA for the RAFT process, sometimes carbon disulfide which is a toxic reagent should be used to prepare CTA. The final RAFT polymers are slightly colored and sometimes a bad odor due to the presence of sulfur atom at the chain end.

2.6 Organotellulium-Mediated Living Radical Polymerization (TERP)

NMP and ATRP proceed via reversible termination mechanism and RAFT proceeds via degenerative transfer mechanism. The mechanism of organotellurium-mediated living radical polymerization (TERP) is reversible termination and degenerative transfer polymerization (Fig. 2.13).

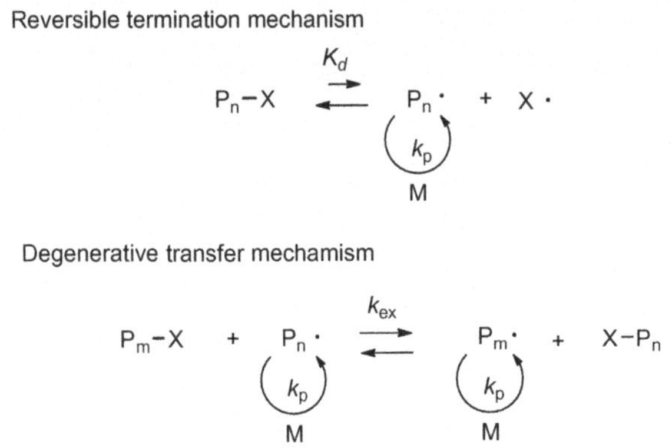

Fig. 2.13 Mechanism of organotellurium-mediated living radical polymerization (TERP)

The k_d value is the first-order rate constant for the activation of the dormant species, and the k_{ex} is the second-order rate constant for the activation via degenerative transfer. For example, the k_d and k_{ex} values using X = *n*-butyl terullium at 60 °C in styrene polymerization are 1×10^{-5} s^{-1} and 3.4×10^3 M^{-1} s^{-1}, respectively [42–44]. Therefore, the degenerative transfer reaction is a dominant mechanism in TERP. PDI of polymers prepared via TERP can be estimated using Eq. 2.16

$$\text{PDI} = M_w / M_n = 1 + \frac{1}{C_{ex}}\left(\frac{2}{\rho} - 1\right) \tag{2.16}$$

where ρ is conversion and C_{ex} is the chain transfer constant (=k_{ex}/k_p), k_p is the rate constant of propagation. A large k_{ex} and high C_{ex} leads to low PDI of the resulting polymer.

Some TERP agents were prepared and purified by vacuum distillation technique (Fig. 2.14). The heteroatom TERP agents are moderately air sensitive. As the synthesis of the TERP agents requires basic conditions, polar functional groups are not compatible.

RAFT polymerization also proceeds by degenerative chain transfer reaction, however, energy profiles in RAFT are different from those of TERP. RAFT polymerization proceeds stepwise with the addition of propagating radicals to the chain transfer agent to form intermediate radical. Subsequent fragmentation generates the propagating radical and macro chain transfer agent. Stability of the intermediate radial in RAFT is strongly affected by the Z group, because it is directly connected to the radical center. On the other hand, TERP proceeds through hypervalent tellurium intermediate or transition state, which forms by reaction of propagating radical with organotellurium dormant species to generate new propagating radical and dormant species. Therefore, chain transfer in TERP virtually proceeds in a concerted manner, and involvement of a long-lived intermediate, which may cause unwanted side

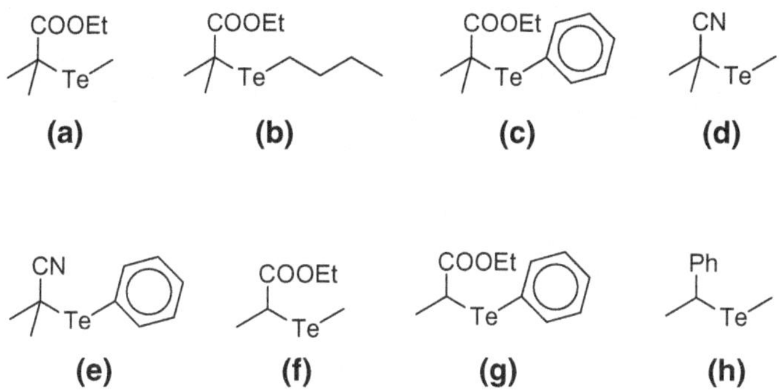

Fig. 2.14 Chemical structures of TERP agents

reactions, is unlikely [45]. The drawback of TERP for practical application is the instability of the TERP agents toward oxygen.

2.7 Block Copolymers

Block copolymers composed of linear or nonlinear different polymer chains are important to prepare soft interfaces. Two main methods of AB diblock copolymer preparation have been developed: (a) coupling of two appropriate end functional polymers and (b) sequential addition of monomers using living polymerization. The sequential addition of monomer is most widely used for the synthesis of block copolymers. The living polymer chains formed the polymerization of the first monomer can efficiently initiate the polymerization of the second monomer.

The covalent coupling of two polymer chains at their ends results in a diblock copolymer (route (a)). There are some examples of coupling of polymers derived by living polymerization have been described. For example, anionic living polystyrene (PS) are coupled with living cationic poly(ethyl vinyl ether) (PEVE) to form corresponding PS-*b*-PEVE [46]. To prepare block copolymers, the formation of metal–ligand interactions of polymers, which are end functionalized by terpyridine units [47]. The prepolymers are coupled by adding ruthenium ions.

The first example of the linear block copolymer was prepared via sequential living anionic polymerization. This technique has emerged as the most reliable and versatile tool for the synthesis of block copolymers [48]. Anionic polymerization proceeds by organometallic sites through nucleophilic reactions in the aprotic solvent. Limitation of anionic polymerization is the demanding experimental conditions require to achieve a living polymerization system and its applicability to several monomers including styrenics, vinyl pyridines, dienes, isoprene, methacrylates, acrylates, ethylene oxide, and octamethyltetracyclosiloxane (D4) (Fig. 2.15) [49]. After the wide use of protective monomers and post-polymerization techniques, the versatility and potential of preparation of block copolymers via anionic polymeriza-

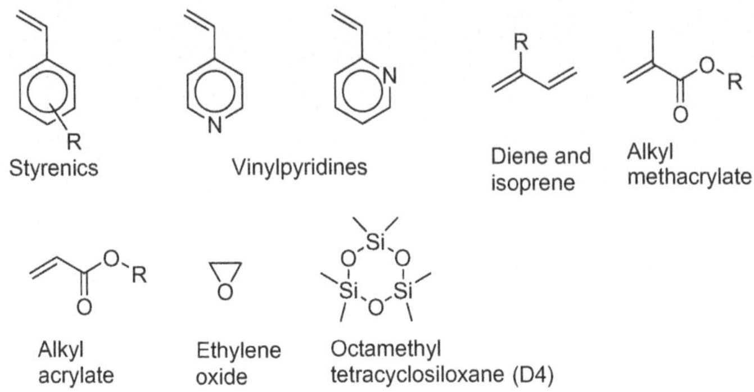

Fig. 2.15 Typical examples of monomers suitable for anionic polymerization

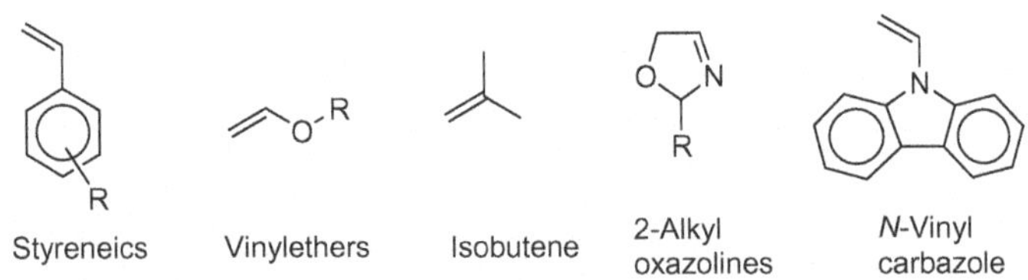

Fig. 2.16 Typical examples of monomers suitable for cationic polymerization

tion have been significantly expected. Diblock copolymers composed of polystyrene and poly(methyl methacrylate) can be prepared [50]. Styrene is polymerized firstly using *sec*-butyl lithium as an initiator. The nucleophilicity of the living PS-lithium should be reduced by reaction with 1,1-diphenyletnylene (DPE) in order to avoid reactions with the carbonyl groups in methyl methacrylate, follows by polymerization of methyl methacrylate.

Cationic polymerization has less control properties, because the inherent instability of carbon atoms is susceptible to chain transfer, isomerization, and termination reactions [2, 51]. Recently novel initiation systems have been developed to promote the living cationic polymerization of a wide variety of monomers, which are those bearing electron-donating groups, including styrenics, vinyl ethers, isobutene, oxazolines, and *N*-vinyl carbazole (Fig. 2.16). A simple synthetic route to block copolymers is to sequentially polymerize monomers with similar reactivity, for example, vinylic monomers of the same family but bearing different substituents. The acetal and trimethylsilyl iodide as the initiator and ZnI_2 activator system can be employed for the preparation of block copolymers of methyl vinyl ether and isobutyl vinyl ether via living cationic polymerization and sequential monomer addition [52].

Compared with ionic polymerizations, free radical polymerizations offer the advantage due to its compatibility with a wide range of monomers, including functional groups. Radical polymerization processes can be performed in an emulsion, suspension, solution using protic and aqueous media, and in bulk. The CLRP techniques have opened up a new window of polymer synthesis. NMP, ATRP, and RAFT are promising polymerization methods to prepare block copolymers with various architectures.

Preparations of block copolymers using TEMPO were reported, for example PS-*block*-polydiene [53], PS-*block*-poly(*n*-butyl acrylate) [54], and so on. The PDIs were rather broad. The conversions of the polymerizations were not quantitative, which were low (<30%) in some case. These observations indicate that the materials were ill-defined. However, α-hydrogen containing alkoxyamines such as DEPN and TIPNO which are called the second generation of persistent nitroxides, enlarged the range of block copolymers accessible by NMP. One of the α-carbon in the second generation alkoxyamines is tertiary. For example acrylates, acrylamides, 1,3-dienes, and acrylonitrile could be readily polymerized with narrow PDIs (<1.1) [55]. The second generation alkoxyamines have enabled NMP to synthesize a broad range of block copolymers. The order in monomer addition is important as demonstrated due to the presence of diblock copolymers which is composed of styrene and acrylate. The successful synthesis of such block copolymers can be achieved starting with the acrylate blocks. If first styrene is polymerized to prepare the diblock copolymer using NMP, the obtained block copolymer contaminated by PS homopolymers.

First, AB block copolymers composed of styrene and methyl acrylate were prepared via ATRP. Many other examples of AB diblock copolymer can be achieved either by reacting the active sites of an isolated macroinitiator or by in situ addition of a second monomer (sequential monomer addition). The second approach sometimes leads to a second block that is a random copolymer of the residual first monomer and the added second monomer. The order of monomer addition is required in ATRP that follows radical stabilities, e.g., acrylonitrile > methacrylate > styrene ≈ acrylates > acrylamides. The monomers which belong to the same family can be obtained well-defined block copolymers with only a small portion of the homopolymers contaminations. In the synthesis of AB diblock copolymers composed of acrylate and methacrylate, a straightforward procedure dictates to initiate with the polymerization of the methacrylate monomer, methyl methacrylate (MMA). PMMA macroinitiator with chlorine terminal can efficiently initiate the polymerization of methyl acrylate (MA), resulting in a well-defined PMMA-*block*-PMA. The opposite order of monomer addition results in a poorly defined diblock copolymer, because the rate of initiation of a PMMA block by a chorine-terminated PMA is slow as compared with the rate of propagation of MMA. One way to improve the efficiency of the block copolymerization is to grow the PMMA block from a bromine-terminated PMA precursor (PMA-Br) in the presence of CuCl (halogen exchange) [56].

The sequential addition of two monomers was mostly used to access block copolymers via RAFT. In contrast to other CRLP methods, the RAFT process uses a macro-CTA instead of a macroinitiator. Block copolymerization of *N,N*-dimethylacrylamide (DMA) and *N,N*-dimethylvinylbenzylamine (DMVBA) via RAFT has been studied using 4,4'-azobis(4-cyanopentanoic acid) as initiator and 4-cyanopentanoic acid dithiobenzoate as a CTA [57]. The block copolymerization reaction proceeds in two steps. First DMVBA macro-CTA was prepared. In a subsequent step, DMA was polymerized, which shows a bimodal GPC distribution. In contrast, starting the block copolymerization from DMA followed by the addition of DMVBA, well-defined block copolymers were obtained. The rate of transfer to the terminal dithioesters carried by a given precursor be higher than the rate of transfer to the dithioesters generated at the end of the growing second block. In other words, it is essential that the first block grown provides a better leaving radical. Therefore, to prepare well-defined diblock copolymer composed of methacrylate and acrylate, the methacrylate monomer should be polymerized first, and then the acrylate monomer should be polymerized.

2.8 Graft Copolymers

Graft copolymers are composed of the main polymer chain and one or more side chains. Sometimes the main chain and side chains are called backbone and branches, respectively. The side chains are chemically connected through covalent bonds to the main chain. The backbone and branches are ordinary homopolymers or copolymers. The branches are often randomly distributed along the backbone. It is known that three strategies to prepare graft copolymers (Fig. 2.17): (a) macromonomer method, (b) grafting onto, and (c) grafting from methods. The most commonly used method to prepare graft copolymers is a macromonomer method [58]. Macromonomer is an oligomeric or polymeric chain bearing a polymerizable end group. Copolymerization of the macromonomer with another monomer yields graft copolymers. For the grafting onto method, ordinary a backbone chain containing functional groups randomly distributed along the backbone. The functional groups are reacted with the chain end of the branched chains. This coupling reaction between the functional backbone and the chain end at branched chains leas to the formation of graft copolymers. In the grafting from method, active sites are generated randomly along the backbone. These sites can initiate polymerization to generate branch chains. Polymerization of the second monomer from the main chain generates graft copolymers.

Homopolymerization of macromonomers affords comb-shaped polymer structures. Copolymerization of macromonomers with another conventional monomer results in the formation of the graft copolymer. Sometimes macromonomer method is called grafting through method. Macromonomer methods are needed for the preparation of macromonomers. Living polymerization techniques including anionic, cationic polymerization, and CRLP can introduce a polymerizable group at the polymer chain end. Among them, CRLP is the most common technique for the synthesis of macromonomers. Before the development of CRLP, the basic method for the introduction of the polymerizable end group involves the use of chain transfer agents such as thiol (SH) compounds [59]. Poly(methyl methacrylate) (PMMA) macromonomers were prepared using thioglycolic acid as a chain transfer agent, followed by reaction with glycidyl methacrylate [60]. CLRP methods led to the synthesis of a wide range of macromonomers [61, 62]. Poly(2-vinyl pyridine) (P2VP) macromonomers are prepared via NMP. The terminal hydroxyl group in the nitroxide group is reacted with methacryloyl chloride to prepare the macromonomers. Copolymerizaiton with *N*-isopropylacrylamide to form graft

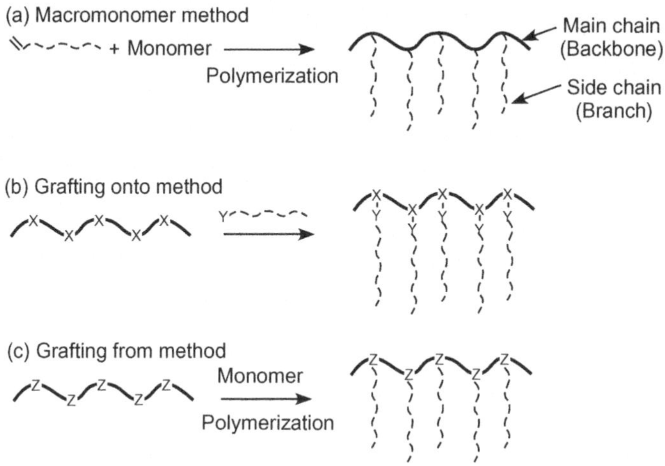

Fig. 2.17 Preparation methods of graft copolymers: **a** macromonomer method, **b** grafting onto method, and **c** grafting from method

copolymers [63]. Poly(dimethylamino ethyl methacrylate) macromonomers were prepared via ATRP [64]. PMMA macromonomers bearing methacrylate at the chain end group are copolymerized with *n*-butyl acrylate and methyl methacrylate (MMA) via ATRP to synthesize the corresponding graft copolymers [65].

Grafting onto methods are to prepare graft copolymers by reaction of polymeric chains bearing functional groups with other polymeric chains bearing active chain ends. The synthesis of polystyrene-*graft*-polyisoprene (PS-*g*-PI) copolymers is prepared using a combination of NMP and anionic polymerization [66]. The PS backbone of PS-*g*-PI is prepared via NMP of styrene and *p*-chloromethyl styrene. The PI living chains prepared via anionic polymerization reacted with the chloromethyl groups to prepare the graft copolymers.

Grafting from methods are to prepare graft copolymers by polymerization of a second monomer from the active site in the backbone. The number of branch chains can be controlled by the number of active sites generated along the backbone. Polypropylene-*graft*-polystyrene (PP-*g*-PS) copolymers are prepared via a combination of NMP and metallocene techniques [67]. The backbone was synthesized via metallocene copolymerization of propylene and a TEMPO-functionalized derivative that contained an α-double bond, through which it is copolymerized. The TEMPO groups are then used for the polymerization of styrene by NMP. Trimethylsilyl-protected 2-hydroxyethyl methacrylate is polymerized via ATRP. The obtained polymer is deprotected to generate pendant hydroxyl groups which are reacted with 2-bromoisobutyl bromide to introduce initiation sites of ATRP. The resulting pendant bromide groups are used as the initiation site for ATRP of styrene and *n*-butyl methacrylate [68].

2.9 Hydrophobically Modified Polyelectrolytes (Random Copolymers)

Amphiphilic polyelectrolytes undergo hydrophobically driven self-association in water to form nanostructures. These polymers are important to design and synthesize novel soft interfaces. The type of nanostructures formed from self-association of amphiphilic polymers depends on whether hydrophobic association occurs within a single polymer chain or between different polymer chains.

A large number of amphiphilic random copolymers of ionic and hydrophobic monomers have been prepared by polymerization. The sequence distribution of the monomer units is an important factor to influence their association behavior [69]. Homogeneous solution polymerization yields copolymers with statistical sequence distributions depending on monomer reactivity ratios, however, heterogeneous polymerization including micelle and emulsion polymerization may give copolymers with blocky sequences [70]. Amphiphilic polyelectrolytes with random sequence distributions can also be prepared by homo- and copolymerization of polymerizable ionic amphiphilies often called as surfmers [71].

Hydrophobic association of random copolymers occurs either in intra- and inter-polymer, the former resulting in a cross-link between polymer chains and the latter in loop within the single polymer chain. Nanostructures formed from amphiphilic random copolymers are greatly different depending on whether inter- or intramolecular hydrophobic association predominantly occurs. When hydrophobes undergo completely intermolecular association, polymer chains are cross-linked, leading to a large increase in solution viscosity due to formation network structure. An infinite network may be formed, as the hydrophobe content is increased to a certain level, forming a gel. When the content of an hydrophobe in a polymer chain is sufficiently low, a flower micelle may be formed, which consists of a hydrophobic core surrounded by ionic loops as theoretically predicted [72, 73]. As the content of the hydrophobe in the polymer is increased, the flower micelles would become unstable because an increasing portion of the hydrophobic core is exposed to water. This would lead to a further collapse into a more compact micelle with a high order structure driven the hydrophobic cores of the flower micelles [74, 75]. However, in most cases, the intermolecular hydrophobic association occurs concurrently with the intermolecular association, where intermolecular bridged flower micelles may be formed. Intra- and intermolecular hydrophobic associations are a powerful function of the content of hydrophobic units in the copolymer. There is a general tendency that intermolecular association is favorable when the content of the hydrophobic unit is low and it becomes unfavorable with increasing the content of hydrophobic units [76–78].

2.10 Gels

Gels including hydrogels are essential materials for the preparation of soft interfaces. Gels are the three-dimensional network structures of polymer and their swollen mattes. They possess both the association properties of solids and the diffusive transport properties of liquid. Their properties vary from viscous liquid to hard solid depending on chemical structures [79].

Gelation refers to the linking of macromolecular chains together which initially leads to progressively larger branched yet soluble polymers depending on the structure and confirmation of the starting material. The mixture of such polydisperse soluble branched polymer is called sol. Continuation of the linking process results in increasing the size of the branched polymer with decreasing solubility. This infinite polymer is called the gel or polymer network and is permeated with finite branched polymers. The transition from a system with a finite branched polymer to infinite molecules is called sol-gel transition (or gelation) and the critical point where gel first appears is called the gel point [80].

Gels can be divided into two main types. (1) Covalent gels, which include bulk elastomers formed form cross-linked materials formed from synthetic polymers and natural polymers. This type of gels is called chemical gels. (2) Non-covalent gels, which are formed from non-covalent interactions as physical gels.

Chemical gels involve formation of covalent bonds and always result in a strong gel. The three main chemical gelation processes include condensation, vulcanization, and addition polymerization. They are called permanent or chemical gels when they are covalently cross-linked networks [81]. They attain an equilibrium swelling state which depends on the polymer–water interaction parameter and the crosslink density [82].

Physical gels are called reversible gels when the networks are held together by molecular entanglements, and/or secondary forces. Physical gels can be subcategorized as strong and weak physical gels. The strong physical gel has strong physical bonds between polymer chains and is effectively permanent at a given set of experimental conditions. Therefore, strong physical gels are likely to chemical gels. Examples of strong physical bonds are lamellar microcrystals, glassy nodules or double and triple helices. Weak physical gels have reversible links formed from temporary associations between chains. These associations have finite lifetimes, breaking and reforming continuously. Examples of weak physical bonds are hydrogen bond, hydrophobic, and ionic interactions.

References

1. Otsu T, Tazaki T (1986) Living radical polymerization in homogeneous system with phenylazotriphenylmethane as a thermal iniferter. Polym Bull 16:277–284. https://doi.org/10.1007/BF00254997
2. Tazaki T, Otsu T (1987) Radical polymerization of methyl methacrylate with methyl 2,2-dimethyl-3,3-diphenyl-3-cyanopropionate as a thermal iniferter. Polym Bull 17:127–134. https://doi.org/10.1007/BF00256877
3. Otsu T, Matsumoto A, Tazaki T (1987) Radical polymerization of methyl methacrylate with some 1,2-disubstituted tetraphenylethanes as thermal iniferters. Polym Bull 17:323–330. https://doi.org/10.1007/BF00955715
4. Qin S-H, Qiu K-Y, Swift G, Westmoreland DG, Wu S (1999) "Living" radical polymerization of methyl methacrylate with diethyl 2,3-dicyano-2,3-diphenylsuccinate as a thermal iniferter. J Polym Sci Part A Polym Chem 37:4610–4615. https://doi.org/10.1002/(SICI)1099-0518(19991215)37:24%3c4610:AID-POLA19%3e3.0.CO;2-O
5. Otsu T, Matunaga T, Doi T, Matsumoto A (1986) Features of living radical polymerization of vinyl monomers in homogeneous system using *N,N*-diethyldithiocarbamate derivatives as photoiniferters. Eur Polym J 31:67–78. https://doi.org/10.1016/0014-3057(94)00122-7
6. Otsu T, Ogawa T, Yamamoto T (1986) Solid-phase block copolymer synthesis by the iniferter technique. Macromolecules 19:2087–2089. https://doi.org/10.1021/ma00161a056
7. Hawker CJ (1994) Molecular weight control by a "living" free-radical polymerization process. J Am Chem Soc 116:11185–11186. https://doi.org/10.1021/ja00103a055
8. Hawker CJ (1997) "Living" free radical polymerization: a unique technique for the preparation of controlled macromolecular architectures. Acc Chem Res 30:373–382. https://doi.org/10.1021/ar960248m
9. Benoit D, Chaplinski V, Braslau R, Hawker CJ (1999) Development of a universal alkoxyamine for "living" free radical polymerizations. J Am Chem Soc 121:3904–3920. https://doi.org/10.1021/ja984013c
10. Catala JM, Bubel F, Oulad Hammouch S (1995) Living radical polymerization: Kinetic results. Macromolecules 28:8441–8443. https://doi.org/10.1021/ma00128a069
11. Fukuda T, Terauchi T, Goto A, Ohno K, Tsujii Y, Miyamoto T, Kobatake S, Yamada B (1996) Mechanisms and kinetics of nitroxide-controlled free radical polymerization. Macromolecules 29:6393–6398. https://doi.org/10.1021/ma960552v
12. Georges MK, Veregin RPN, Kazmier PM, Hamer GK (1993) Narrow molecular weight resins by a free-radical polymerization process. Macromolecules 26:2987–2988. https://doi.org/10.1021/ma00063a054
13. Hawker CJ, Bosman AW, Harth E (2001) New polymer synthesis by nitroxide mediated living radical polymerizations. Chem Rev 101:3661–3688. https://doi.org/10.1021/cr990119u
14. Benoit D, Grimaldi S, Finet J-P, Tordo P, Fontanille M, Gnanou Y (1998) Controlled/living free-radical polymerization of styrene and *n*-butyl acrylate in the presence of a novel asymmetric nitroxyl radical. In: Matyjaszewski K (ed) Controlled radical polymerization. ACS Symposium Series vol 685. American Chemical Society
15. Hawker CJ, Barclay GG, Orellana A, Dao J, Devonport W (1996) Initiating systems for nitroxide-mediated "living" free radical polymerizations: synthesis and evaluation. Macromolecules 29:5245–5254. https://doi.org/10.1021/ma951905d
16. Wang J-S, Matyjaszewski K (1995) Controlled/"living" radical polymerization. Halogen atom transfer radical polymerization promoted by a Cu(I)/Cu(II) redox process. Macromolecules 28:7901–7910. https://doi.org/10.1021/ma00127a042
17. Wang J-S, Matyjaszewski K (1995) Controlled/"living" radical polymerization. atom transfer radical polymerization in the presence of transition-metal complexes. J Am Chem Soc 117:5614–5615. https://doi.org/10.1021/ja00125a035
18. Granel C, Dubois P, Jérôme R, Teyssié P (1996) Controlled radical polymerization of methacrylic monomers in the presence of a bis(ortho-chelated) arylnickel (II) complex

and different activated alkyl halides. Macromolecules 29:8576–8582. https://doi.org/10.1021/ma9608380

19. Uegaki H, Kotani Y, Kamigaito M, Sawamoto M (1997) Nickel-mediated living radical polymerization of methyl methacrylate. Macromolecules 30:2249–2253. https://doi.org/10.1021/ma961367k
20. Kato M, Kamigaito M, Sawamoto M, Higashimura T (1995) Polymerization of methyl methacrylate with the carbon tetrachloride/dichlorotris–(triphenylphosphine)ruthenium (II)/methylaluminum bis(2,6-di-tert-butylphenoxide) initiating dystem: Possibility of living radical polymerization. Macromolecules 28:1721–1723. https://doi.org/10.1021/ma00109a056
21. Haddleton DM, Jasieczek CB, Hannon MJ, Shooter AJ (1997) Atom transfer radical polymerization of methyl methacrylate initiated by alkyl bromide and 2-pyridinecarbaldehyde imine copper (I) complexes. Macromolecules 30:2190–2193. https://doi.org/10.1021/ma961074r
22. Tang W, Kwak Y, Braunecker W, Tsarevsky NV, Coote ML, Matyjaszewski K (2008) Understanding atom transfer radical polymerization: effect of ligand and initiator structures on the equilibrium constants. J Am Chem Soc 130:10702–10713. https://doi.org/10.1021/ja802290a
23. Matyjaszewski K, Patten TE, Xia J (1997) Controlled/"living" radical polymerization. Kinetics of the homogeneous atom transfer radical polymerization of styrene. J Am Chem Soc 119:674–680. https://doi.org/10.1021/ja963361g
24. Litvinenko G, Mueller AHE (1997) General kinetic analysis and comparison of molecular weight distributions for various mechanisms of activity exchange in living polymerizations. Macromolecules 30:1253–1266. https://doi.org/10.1021/ma960945u
25. Tang W, Matyjaszewski K (2007) Effects of initiator structure on activation rate constants in ATRP. Macromolecules 40:1858–1863. https://doi.org/10.1021/ma062897b
26. Tang W, Matyjaszewski K (2006) Effect of ligand structure on activation rate constants in ATRP. Macromolecules 39:4953–4959. https://doi.org/10.1021/ma0609634
27. Chiefari J, Chong YK, Ercole F, Krstina J, Jeffery J, Le TPT, Mayadunne RTA, Meijs GF, Moad CL, Moad G, Rizzardo E, Thang SH (1998) Living free-radical polymerization by reversible addition—fragmentation chain transfer: the RAFT process. Macromolecules 31:5559–5562. https://doi.org/10.1021/ma9804951
28. Barner-Kowollik C, Buback M, Charleux B, Coote ML, Drache M, Fukuda T, Goto A, Klumperman B, Lowe AB, Mcleary JB, Moad G, Monteiro MJ, Sanderson RD, Tonge MP, Vana P (2006) Mechanism and kinetics of dithiobenzoate-mediated RAFT polymerization. I. The current situation. J Polym Sci Part A Polym Chem 44:5809–5831. https://doi.org/10.1002/pola.21589
29. Ting SRS, Davis TP, Zetterlund PB (2011) Retardation in RAFT polymerization: Does cross-termination occur with short radicals only? Macromolecules 44:4187–4193. https://doi.org/10.1021/ma200831f
30. Junkers T (2011) RAFT kinetics revisited: revival of the RAFT debate. J Polym Sci Part A Polym Chem 49:4154–4163. https://doi.org/10.1002/pola.24857
31. Chong YK, Krstina J, Le TPT, Moad G, Postma A, Rizzardo E, Thang SH (2003) Thiocarbonylthio compounds [S=C(Ph)S-R] in free radical polymerization with reversible addition-fragmentation chain transfer (RAFT polymerization). Role of the free-radical leaving group (R). Macromolecules 36:2256–2272. https://doi.org/10.1021/ma020882h
32. Muller AHE, Zhuang R, Yan D, Litvinenko G (1995) Kinetic analysis of "living" polymerization processes exhibiting slow equilibria. 1. Degenerative transfer (direct activity exchange between active and "dormant" species). Application to group transfer polymerization. Macromolecules 28:4326–4333. https://doi.org/10.1021/ma00116a040
33. Barner-Kowollik C, Quinn JF, Morsley DR, Davis TP (2001) Modeling the reversible addition–fragmentation chain transfer process in cumyl dithiobenzoate-mediated styrene homopolymerizations: Assessing rate coefficients for the addition–fragmentation equilibrium. J Polym Sci Part A Polym Chem 39:1353–1365. https://doi.org/10.1002/pola.1112
34. Chernikova EV, Tarasenko AV, Garina ES, Golubev VB (2006) Controlled radical polymerization of styrene mediated by dithiobenzoates as reversible addition-fragmentation chain-transfer agents. Polym Sci Ser A 48:1046–1057. https://doi.org/10.1134/S0965545X06100051

35. Chiefari J, Mayadunne RTA, Moad CL, Moad G, Rizzardo E, Postma A, Thang SH (2003) Thiocarbonylthio compounds (SC(Z)S–R) in free radical polymerization with reversible addition-fragmentation chain transfer (RAFT polymerization). Effect of the activating group Z. Macromolecules 36:2273–2283. https://doi.org/10.1021/ma020883+
36. Postma A, Davis TP, Evans RA, Li G, Moad G, O'Shea M (2006) Synthesis of well-defined polystyrene with primary amine end groups through the use of phthalimido-functional RAFT agents. Macromolecules 39:5293–5306. https://doi.org/10.1021/ma060245h
37. Adamy M, van Herk AM, Destarac M, Monteiro MJ (2003) Influence of the chemical structure of MADIX agents on the RAFT polymerization of styrene. Macromolecules 36:2293–2301. https://doi.org/10.1021/ma025721s
38. Theis A, Davis TP, Stenzel MH, Barner-Kowollik C (2006) Probing the reaction kinetics of vinyl acetate free radical polymerization via living free radical polymerization (MADIX). Polymer 47:999–1010. https://doi.org/10.1016/j.polymer.2005.12.054
39. Chernikova EV, Terpugova PS, Garina ES, Golubev VB (2007) Controlled radical polymerization of styrene and *n*-butyl acrylate mediated by trithiocarbonates. Polym Sci Ser A 49:108–119. https://doi.org/10.1134/S0965545X07020022
40. Moad G, Rizzardo E, Thang SH (2005) Living radical polymerization by the RAFT process. Aust J Chem 58:379–410. https://doi.org/10.1071/CH05072
41. Moad G, Rizzardo E, Thang SH (2008) Radical addition–fragmentation chemistry in polymer synthesis. Polymer 49:1079–1131. https://doi.org/10.1016/j.polymer.2007.11.020
42. Goto A, Kwak Y, Fukuda T, Yamago S, Iida K, Nakajima M, Yoshida J (2003) Mechanism-based invention of high-speed living radical polymerization using organotellurium compounds and azo-initiators. J Am Chem Soc 125:8720–8721. https://doi.org/10.1021/ja035464m
43. Kayahara E, Yamago S, Kwak Y, Goto A, Fukuda T (2008) Optimization of organotellurium transfer agents for highly controlled living radical polymerization. Macromolecules 41:527–529. https://doi.org/10.1021/ma7023717
44. Kwak Y, Goto A, Fukuda T, Kobayashi Y, Yamago S (2006) A systematic study on activation processes in organotellurium-mediated living radical polymerizations of styrene, methyl methacrylate, methyl acrylate, and vinyl acetate. Macromolecules 39:4671–4679. https://doi.org/10.1021/ma060295m
45. Schiesser CH, Smart BA, Tran TA (1995) An ab initio study of some free-radical homolytic substitution reactions at halogen. Tetrahedron 51:3327–3338. https://doi.org/10.1016/0040-4020(95)00054-C
46. Creuts S, Vanooren C, Jérôme R, Teyssié P (1994) Coupling of living polystyryllithium anions and living polyethyl vinyl ether cations. Polm Bull 33:21–28. https://doi.org/10.1007/BF00313469
47. Gohy JF, Lohmeijer BGG, Schubert US (2003) From supramolecular block copolymers to advanced nano-objects. Chem Eur J 9:3472–3479. https://doi.org/10.1002/chem.200204640
48. Hsieh HL, Quirk RP (1996) Anionic polymerization: principles and practical applications. Marcel Dekker, New York
49. Hadjichristidis N, Iatrou H, Pispas S, Pitsikalis M (2000) Anionic polymerization: high vacuum techniques. J Polym Sci Part A Polym Chem 38:3211–3234. https://doi.org/10.1002/1099-0518(20000915)38:18%3c3211:AID-POLA10%3e3.0.CO;2-L
50. McGrath J (ed) (1981) Anionic polymerization. In: ACS Symposium Series 166. American Chemical Society
51. Matyjaszewski, K. (ed) (1996) Cationic polymerization. Mechanisms, synthesis and applications. Marcel Dekker, New York
52. Verdonck B, Goethals EJ, Du Prez FE (2003) Radical cation, dimer radical cation and neutral radical of 2,3-dihydropyran–possible initiators of its polymerisation? Macromol Chem Phys 204:2090–2098. https://doi.org/10.1002/macp.200350068
53. Georges MK, Hamer GK, Listigovers NA (1998) Block copolymer synthesis by a nitroxide-mediated living free radical polymerization process. Macromolecules 31:9087–9089. https://doi.org/10.1021/ma970758b

54. Chong BYK, Ercole F, Moad G, Rizzardo E, Thang SH (1999) Imidazolidinone nitroxide-mediated polymerization. Macromolecules 32:6895–6903. https://doi.org/10.1021/ma9904868
55. Benoit D, Harth E, Fox P, Waymouth RM, Hawker CJ (2000) Accurate structural control and block formation in the living polymerization of 1,3-dienes by nitroxide-mediated procedures. Macromolecules 33:363–370. https://doi.org/10.1021/ma991187l
56. Shipp DA, Wang JL, Matyjaszewski K (1998) Synthesis of acrylate and methacrylate block copolymers using atom transfer radical polymerization. Macromolecules 31:8005–8008. https://doi.org/10.1021/ma981033q
57. Sumerlin BS, Lowe AB, Thomas DB, Convertine AJ, Donovan MS, McCormick CL (2004) Aqueous solution properties of pH-responsive AB diblock acrylamido-styrenic copolymers synthesized via aqueous reversible addition-fragmentation chain transfer. J Polym Sci Part A Polym Chem 42:1724–1734. https://doi.org/10.1002/pola.11069
58. Hadjichristidis N, Pitsikalis M, Iatrou H, Pispas S (2003) The strength of the macromonomer strategy for complex macromolecular architecture: Molecular characterization, properties and applications of polymacromonomers. Macromol Rapid Commun 24:979–1013. https://doi.org/10.1002/marc.200300050
59. Odian G (1981) Principles of polymerization, 2nd edn. Wiley Interscience, New York
60. Yamashita Y, Ito K, Mizuno H, Okada K (1982) Polym J 14:255–260. https://doi.org/10.1295/polymj.14.255
61. Malmstrom EE, Hawker CJ (1998) Macromolecular engineering via 'living' free radical polymerizations. Macromol Chem Phys 199:923–935. https://doi.org/10.1002/(SICI)1521-3935(19980601)199:6%3c923:AID-MACP923%3e3.0.CO;2-Y
62. Matyjaszewski K (ed) (1997) Controlled radical polymerization. ACS symposium series No. 685
63. Wohlrab S, Kuckling D (2001) Multisensitive polymers based on 2-vinylpyridine and *N*-isopropylacrylamide. J Polym Sci A Polym Chem 39:3797–3804. https://doi.org/10.1002/pola.10026
64. Zeng F, Shen Y, Zhu S, Pelton R (2000) Synthesis and characterization of comb-branched polyelectrolytes. 1. Preparation of cationic macromonomer of 2-(dimethylamino)ethyl methacrylate by atom transfer radical polymerization. Macromolecules 33:1628–1635. https://doi.org/10.1021/ma991654z
65. Roos SG, Mu1ller AHE, Matyjaszewski K (1999) Copolymerization of n-butyl acrylate with methyl methacrylate and PMMA macromonomers: comparison of reactivity ratios in conventional and atom transfer radical copolymerization. Macromolecules 32:8331–8335. https://doi.org/10.1021/ma9819337
66. Tsoukatos T, Pispas S, Hadjichristidis N (2000) Complex macromolecular architectures by combining TEMPO living free radical and anionic polymerization. Macromolecules 33:9504–9511. https://doi.org/10.1021/ma0011946
67. Stehling UM, Malmstrom EE, Waymouth RM, Hawker CJ (1998) Synthesis of poly(olefin) graft copolymers by a combination of metallocene and "living" free radical polymerization techniques. Macromolecules 31:4396–4398. https://doi.org/10.1021/ma980141+
68. Beers KL, Gaynor SG, Matyjaszewski K, Sheiko SS, Moeller M (1998) The synthesis of densely grafted copolymers by atom transfer radical polymerization. Macromolecules 31:9413–9415. https://doi.org/10.1021/ma981402i
69. Cheng Y, McCormick CL (1993) Water-soluble copolymers. 49. Effect of the distribution of the hydrophobic cationic monomer dimethyldodecyl(2-acrylamidoethyl) ammonium bromide on the solution behavior of associating acrylamide copolymers. Macromolecules 26:6121–6126. https://doi.org/10.1021/ma00074a038
70. Branham KD, Shafer GS, Hoyle CE, McCormick CL (1995) Water-soluble copolymers. 61. Microstructural investigation of pyrenesulfonamide-labeled polyelectrolytes. Variation of label proximity utilizing micellar polymerization. Macromolecules 28:6175–6182. https://doi.org/10.1021/ma00122a026

71. Laschewsky A (1995) Molecular concepts, self-organisation and properties of polysoaps. Adv Polym Sci 124:1–86. https://doi.org/10.1007/BFb0025228
72. Halperin A (1991) On the collapse of multiblock copolymers. Macromolecules 24:1418–1419. https://doi.org/10.1021/ma00006a033
73. Semenov AN, Joanny J-F, Khokhlov AR (1995) Associating polymers: equilibrium and linear viscoelasticity. Macromolecules 28:1066–1075. https://doi.org/10.1021/ma00108a038
74. Yamamoto H, Morishima Y (1999) Effect of hydrophobe content on intra- and inter-polymer self-associations of hydrophobically modified poly(sodium 2-(acrylamido)-2-methylpropanesulfonate) in water. Macromolecules 32:7469–7475. https://doi.org/10.1021/ma9907791
75. Yamamoto H, Mizusaki M, Yoda K, Morishima Y (1998) Fluorescence studies of hydrophobic association of random copolymers of sodium 2-(acrylamido)-2-methylpropanesulfonate and *N*-dodecylmethacrylamide in water. Macromolecules 31:3588–3594. https://doi.org/10.1021/ma980022u
76. McCormick CL, Chang Y (1994) Water-soluble copolymers. 58. Associative interactions and photophysical behavior of amphiphilic terpolymers prepared by modification of maleic anhydride/ethyl vinyl ether copolymers. Macromolecules 27:2151–2158. https://doi.org/10.1021/ma00086a026
77. Chang Y, McCormick CL (1994) Water-soluble copolymers: 57. Amphiphilic cyclocopolymers of diallylalkoxybenzyl-methylammonium chloride and diallyl-dimethylammonium chloride. Polymer 35:3503–3512. https://doi.org/10.1016/0032-3861(94)90916-4
78. Petit F, Iliopoulos I, Audebert R, Szönyi S (1997) Associating polyelectrolytes with perfluoroalkyl side chains: aggregation in aqueous solution, association with surfactants, and comparison with hydrogenated analogues. Langmuir 13:4229–4233. https://doi.org/10.1021/la970003y
79. Sekhon SS (2003) Conductivity behaviour of polymer gel electrolytes: role of polymer. Bull Mater Sci 26:321–328
80. Rubinstein M, Colby RH (2003) Polymer physics. Oxford University Press, Oxford
81. Hennink WE, Nostrum CFV (2002) Novel crosslinking methods to design hydrogels. Adv Drug Deliv Rev 54:13–36. https://doi.org/10.1016/S0169-409X(01)00240-X
82. Rosiak JM, Yoshii F (1999) Hydrogels and their medical applications. Nucl Instrum Methods Phys Res B 151:56–64. https://doi.org/10.1016/S0168-583X(99)00118-4

Chapter 3
Nano- and Micro-technology of Soft Interface

Yoshiko Miura and Keitaro Yoshimoto

3.1 Supramolecular Formation

Supramolecules formation is the essential concept to control the soft materials, and the basics of bottom-up technology. The various and well-defined supramolecules are found in nature such as lipid bilayer and DNA hybridization. The well-defined supramolecules have been investigated, learned from nature. In order to fabricate the soft matter by supramolecule, the control of nanostructure orientation and hierarchical assembly are required.

3.1.1 Supramolecular Formation with Biomacromolecules

In nature, the molecules are self-assembled into the precise structures like lipid bilayer, DNA hybridization, and protein folding. The self-assembled molecules autonomously form the macroscopic structure by hierarchical interactions. Among the biomacromolecules, DNA are well known as regularly self-assembled structure, i.e., double-strand helical structure. The self-assembling structure of DNA is accomplished by the complementary hydrogen bonding formation between nucleic acids (adenine (A) and thymine (T), cytosine (C) and guanine (G)). On the contrary, it is possible to design the nanomaterials with synthetic DNA. Seeman first reported the designed nanostructure of synthetic DNA where he reported the 2D-ordered structure of DNA (Fig. 3.1) [1]. Various groups have reported the regular nanostructure

Y. Miura (✉)
Department of Chemical Engineering,, Kyushu University, Fukuoka, Japan
e-mail: miuray@chem-eng.kyushu-u.ac.jp

K. Yoshimoto
Graduate School of Arts and Science, The University of Tokyo, Tokyo, Japan

M. Maeda et al. (eds.), *Molecular Soft-Interface Science*,
https://doi.org/10.1007/978-4-431-56877-3_3

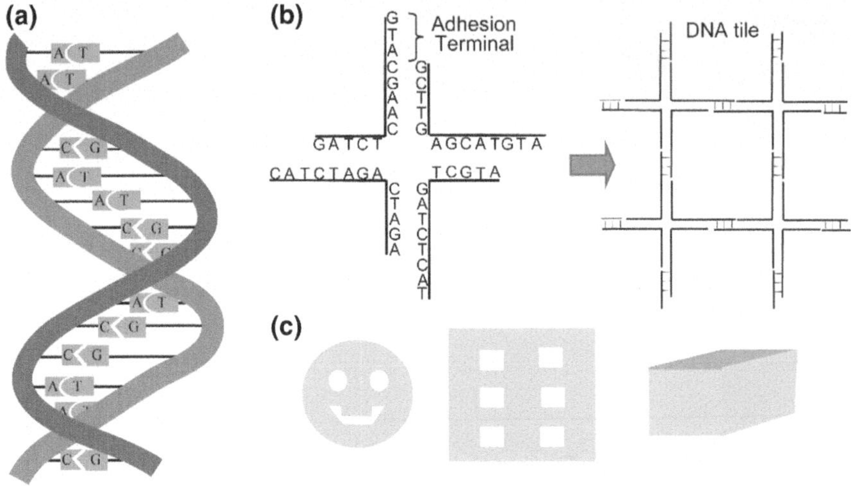

Fig. 3.1 DNA supramolecules. **a** DNA double strand, **b** the concept of DNA nanostructure formation, and **c** the example of DNA nanostructures

formation of the artificial DNA sequences, Recently, Rothemund reported the DNA origami to construct the ordered nanostructure [2].

Peptide supramolecules were also investigated. Peptides form the regular secondary structure like α-helices and β-sheets in nature [3]. The synthetic peptides with secondary structure are semi-empirically designed by de novo design. The secondary structures are affected by the kinds of amino acid residues and the hydrophobic interactions. The helix bundle with the synthetic peptides have been reported as an artificial protein. The functional helical peptides have been reported to show the activities such as the enzyme-mimic catalyst, ion channel, and antimicrobials [4]. The synthetic β-sheet peptides have been also reported, which form the gels and nanowires [5]. Deming et al. prepared the various polypeptides with α-helices, β-strand and random coil structures via metal-catalyzed NCA polymerization. They reported that peptides formed hydrogels, and that the rheological properties were correlated the peptide secondary structure [6].

3.1.2 Supramolecular Nanomaterials

Well–ordered structures are able to be prepared by self-assembling processes, with the biomacromoelcuels and synthetic molecules. Lehn suggested the self-assembled molecules by molecular interactions are “Supramolecules” [7]. The supramolecular materials are constructed not by the covalent bond, but by the molecular interactions such as hydrogen bonding, electrostatic interaction, metal complex formation, charge-transfer complex, and solvent–phobic interactions. The molecular interac-

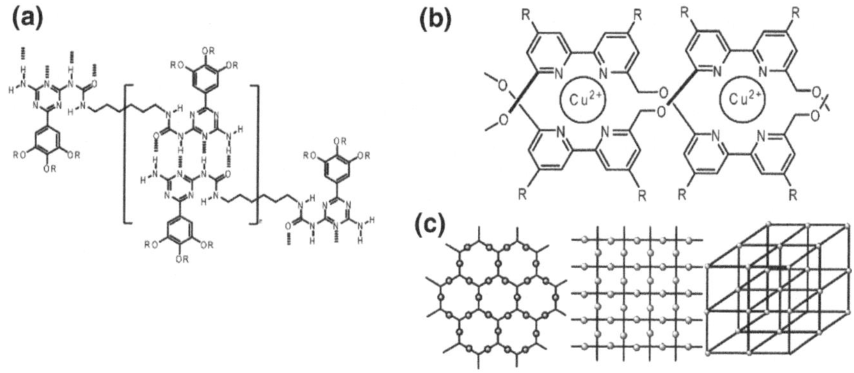

Fig. 3.2 Supramolecular nanomaterials. **a** Supramolecular polymer with hydrogen bonding, **b** supramolecular polymer with metal complexation (helicate), and **c** the examples of supramolecules with metal complexations of MOF

tions are basically much weaker than the covalent bonding, but the multiple molecular interactions are strong enough to form the molecular assembly.

Among the various molecular interactions, hydrogen bonding is useful and provides the selectivity and directionality (Fig. 3.2a). Functional groups that form multiple hydrogen bonding are useful building blocks of supramolecules. For example, melamine and barbituric acid are able to form six hydrogen bonding in a molecule. Whitesides et al. reported that barbituric acid and melamine form the rose-like supramolecular crystal [8]. Ringsdorf et al. reported that the synthetic lipid having melamine formed 2D crystal via hydrogen bonding with such as cyanuric acid [9]. Meijer et al. reported the chiral self-assembly of liquid crystal by hydrogen bonding [10].

Metal complexation is also a useful interaction to form supramolecules (Fig. 3.2b). In the early studies, Lehn reported the ordered structure of bipyridine and copper, called "helicate", which showed the ordered structure like DNA double-strand [11]. Fujita et al. reported various ordered structures like cages with metal complexation [12]. Metal complexation even forms the large three-dimensional porous materials, which are called metal–organic framework (MOF) [13]. MOF forms the regular porous nanostructure via metal complexation, where the mechanical strength of MOF is strong enough. MOF provides the ordered nanospaces which is able to be applied for the various functional materials such as gas adsorption, catalyst and water purification.

Not only utilizing the molecular interactions, but also the solvent effects are important. The solvophobic interactions play important roles in the formation of supramolecules. It is well known that the amphiphilic molecules are self-assembled by hydrophobic effect. For example, the lipids spontaneously form the self-assembling structure like bilayer (liposomes), micelles and bicelles (Fig. 3.3) [14]. Most of the self-assembling of biomacromolecules are affected by the hydrophobic effect, and the hydrophobic effect is of course useful for the construction of

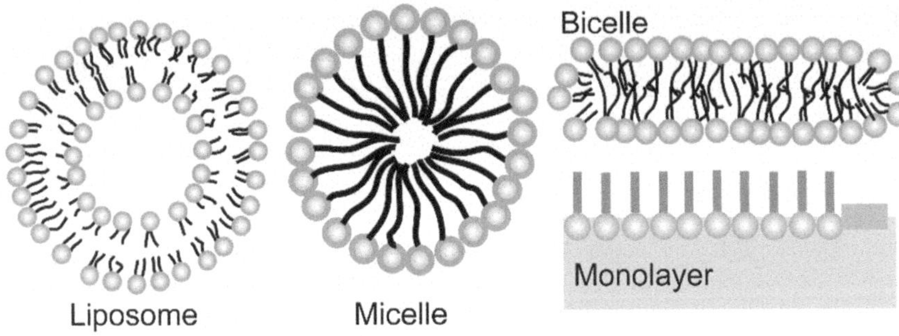

Fig. 3.3 Supramolecular formations of amphiphiles by hydrophobic interactions

the synthetic supramolecules. Kunitake et al. first reported the bilayer formation with the synthetic lipids [15]. Now, various amphiphiles are reported to form the supramolecules by hydrophobic interaction. The amphiphilic block polymers have been reported to form micelles in the aqueous solutions, which are applied to the drug delivery systems [16].

The combination of the molecular interaction and the solvophobic interactions are effective to prepare the highly ordered and macroscopic materials. The lipids with a functional group such as hydrogen bonding motif and complexation ligand are able to form the highly ordered and large supramolecules. For example, Ringsdorf reported that the lipid–melamine conjugate formed 2D crystal at the air–water interface. Kimizuka et al. reported that metal nanowire by the ligand–lipid conjugates [17].

Supramolecular is one of the important methods to fabricate the soft materials. In order to obtain the highly ordered supramolecules, the appropriate molecular interaction (hydrogen bonding) is important, and the solvent assists the supramolecular formation.

3.1.3 Langmuir and Langmuir–Blodgett Membranes

Air–water interface is the useful reaction field for molecular assemblies. Generally, the amphiphilic surfactant forms the stable monolayer at the air–water interface. The balance of amphiphilicity should affect the monolayer formation (Fig. 3.4). The surfactants having the following surface tension are able to form the stable monolayer (Fig. 3.5):

$$\gamma_w \geq \gamma_o \cos\theta + \gamma_{wo} \cos\theta'$$

where γ_{w}, γ_{o}, and γ_{wo} are surface tension of water, surface tension of the oil (surfactant), and surface tension of the water–oil interface. θ and θ’ represent the angle of the tangent of each surface tension.

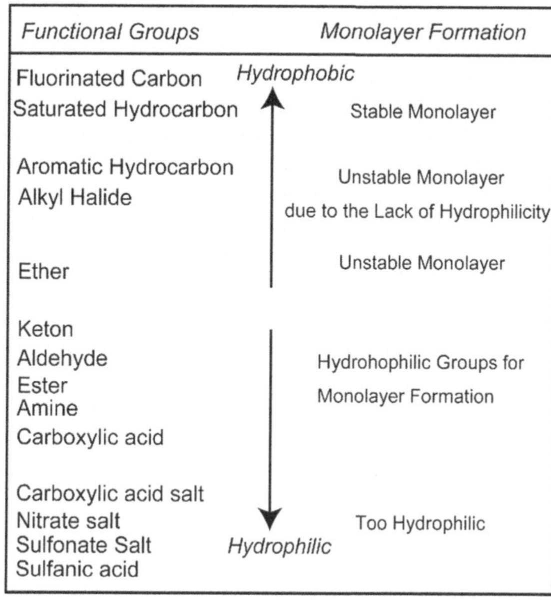

Fig. 3.4 The ability of monolayer formation at the air–water interface

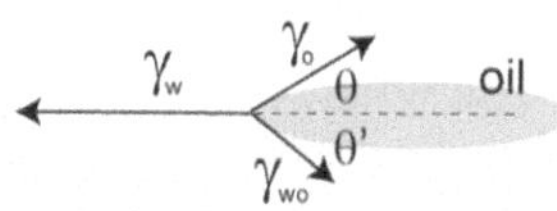

Fig. 3.5 The condition of the molecules for monolayer formation at the air–water interface

The monolayer at the air–water interface is compressed and accumulated with a special apparatus of the trough which is composed of water bath, moving barrier, and pressure meter. The physical–chemical properties are shown in π-A isotherms, where π and A are surface pressure at the water interface and area per molecule, respectively. π -A isotherm is a kind of 2D phase diagram, which resembles to PV isotherm in 3D space. Figure 3.6 presents a π -A isotherm of a surfactant. The area (A) decrease correlates with the pressure increase.

The monolayer states are defined as "gas", "liquid" and "solid" phases. In the gas state, A is large, and the interaction between molecules is low. In the smaller A, the molecule interacts with the adjacent molecules like liquid and solid phases, and the surface pressure increases. In the case of simple lipids such as stearic acid, the gas–liquid transition and the liquid–solid transition occurs at area per molecules of around 24 $Å^2$ and 22 $Å^2$. Considering the area per molecules in the solid state, the molecules form the condensed monolayer of 2D crystal.

The monolayer at the air–water interface can be transferred to the solid substrate. The multilayers of the surfactant are able to be prepared by the repetition of monolayer transfer. This multilayer is called Langmuir–Blodgett film (LB film), and the monolayer at the air–water interface is called Langmuir film. The deposition of the monolayer to the substrate is divided into vertical deposition and the horizontal

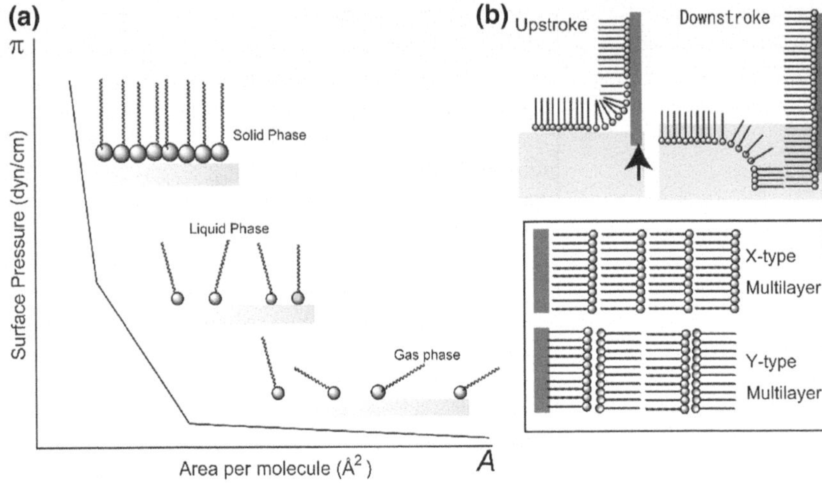

Fig. 3.6 Schematic image of Langmuir and Langmuir–Blodgett membranes. **a** A π-A isotherm with lipid phases and **b** deposition of a monolayer from the air–water interface to a vertical plate

deposition. In the vertical deposition, the monolayer is able to be transferred during emersion (upstroke) and immersion (downstroke). During the emersion, a monolayer is transferred with the head hydrophilic group attached to the substrate, and during immersion, the layer is transferred by the opposite way. The transfer ratio is affected by the interaction between the monolayer and the substrate. The multilayer orientation are three ways by the combination of the deposition method. The merit of the LB layer is the facile control of layer thickness and orientation.

While the amphiphilicity is important in the case of small molecules, the amphiphilicity is not always critical in the case of Langmuir layer of polymers. Various hydrophobic polymers, such as poly(methyl methacrylate) (PMMA), polystyrene (PSt), alkylated cellulose, and alkylated polyacrylamide, were reported to form monolayer and LB layer [18]. Hydrophilic polymers are difficult to form monolayer and LB layer, but block polymers with hydrophobic segment are able to form those layers. In the LB layer of the small molecules, the molecules form the densely packed structure, but the defects from the energy defects are found. On the other hand, in the LB layer of polymers, the molecular packings are not tight like the small molecules, but the LB layers are uniform.

3.1.4 Self-assembled Monolayer (SAM)

The amphiphilic molecules can form the ordered and well-packed LB layer. However, the molecules without the appropriate amphiphilicity are not suitable to form the LB layer. In addition, the special apparatus of LB trough is required.

The self-assembled monolayers (SAMs) are spontaneously formed monolayers on the substrate based on the "self-assembling properties" (Fig. 3.7) [19]. The alkanethiol SAM is a densely packed structure, similar to the Langmuir monolayer. The strong interaction to the substrate and the lateral force between the molecules are important in the SAM formation. The combination of the functional group and the substrate are divided into two: alkanethiol and precious metals (Au Pt and Ag), alkoxysilane, and hydroxyl group of oxidized layer. SAMs are prepared by a simple procedure of substrate immersion into the solvent, and the molecules are spontaneously assembled to form a densely packed monolayer. The orientation and the packing of the SAMs depend on the self-assembling properties such as the molecular structure, the concentration of the molecules, and the solvent system. The alkanethiols or alkoxysilane with longer alkyl chain (>C8) form the densely packed and oriented monolayer on the substrate. The SAMs of other organic molecules such as peptide, DNA, cyclodextrin, and oligoethylene glycol-alkanethiol were reported. The SAMs are the facile technique to modify the surface of the substrate, which is utilized for the preparation of electronic devices, biosensors, biomaterials, and surface modifications.

On the other hand, the monolayer formation is easy to be accomplished, but the preparation of multilayer is difficult. Sagiv et al. reported the primary challenge of multilayer SAMs [20]. They prepared the SAM of alkylsilane with a vinyl group that was decomposed to hydroxyl group, and deposited another alkylsilane layer on the hydroxyl group. The successive formation of SAMs leads to the multilayer formation, but the defects of the monolayer significantly affect the multilayer.

3.1.5 Alternate Layer-by-Layer Assembly (LBL)

The alternate layer-by-layer assembly is a method to form multilayer based on the self-assembling properties of molecules. The layer-by-layer (LBL) technique was first reported by Decher in the 1990s [21]. The building blocks of LBL are usually polyelectrolyte, and the multilayers are formed by the interaction between the poly-

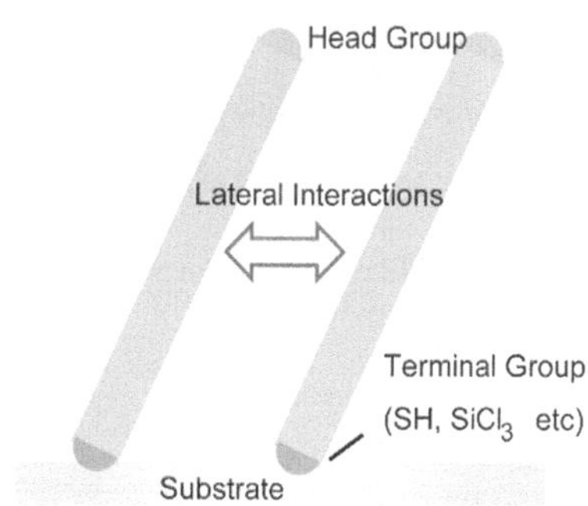

Fig. 3.7 Schematic illustration of the self-assembled monolayer (SAM)

mers. In the case of polyelectrolyte, a functional group with a charge (positive or negative) is immobilized, and the substrate is alternately dipped into the oppositely charged polyelectrolyte solution. Polymer layer forms like + − + − + − (Fig. 3.8).

The amount of polymer adsorption is larger than that for charge neutralization, because the polymer adsorption occurs due to not only the electrostatic interaction but also the molecular interactions. The excess charge on the periphery induces the next polymer layer adsorption. It has been reported that the polymer diffusion into the LBL films occurs and the ion exchanges occurs, suggesting the LBL layer formation is not the simple electrostatic adsorption. The LBL layer structure depends on the building block polymer structures, and is affected by the pH of solution, salt concentration, and charges on the polymer. The small molecules without large molecular interaction do not form LBL layer due to the lack of mutual polymer interactions. In other words, various interaction can be applied to LBL such as hydrogen bonding and stereo complexation. Both synthetic and natural polymers are able to be utilized: Kunitake group reported the protein multilayer formation by LBL technique [22]. Hammond group reported the multilayer formation by hydrogen bonding [23]. Akashi et al. reported the multilayer formation by stereo complexation [24]. Recently the LBL technique is applied to the preparation of various materials, such as surface modification (substrate and particles), electronic materials, separation membranes and biomaterials including 3D cell culture.

LBL technique is a unique and promising method for multilayer formation. The materials properties by LBL are controlled by the interaction between polymers, the kind of polymers, the polymer conformation and the energy dissipation of the polymers. The LBL technique doesn't need the special apparatus, and provide the multilayer by low cost and low energy processes.

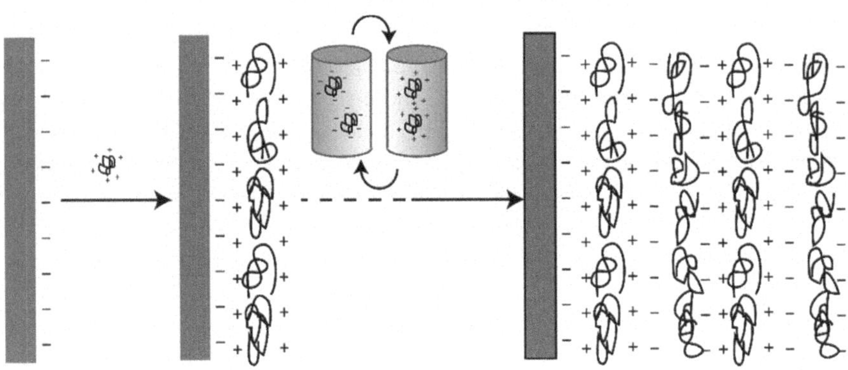

Fig. 3.8 Multilayer formation by layer-by-layer assembly

3.1.6 Polymers with Special Structure

Polymer's physical theories are summarized with coiled-coil polymers in a structured and consistent way by Flory. However, polymers with special structures have different physical properties. Those polymers show the unique properties The branched polymers and polyrotaxanes are mentioned below.

3.1.6.1 Dendrimers

Polymers usually have the polydispersities, even though the living polymers have narrow polydispersities, and form the ordered uniform nanostructure with one molecule. Since dendrimers are regularly branched molecules from core to the terminal, the properties of core and terminal are different from each other.

The synthetic method of dendrimers is classified into two, namely divergent and convergent methods (Fig. 3.9). In 1985, Tomalia reported the synthesis of polyamidoamine (PAMAM) dendrimers, where the reaction involved the Michael addition to methyl acrylate and the amidation [25]. In the divergent method, the dendrimers are synthesized from the core to the periphery (Fig. 3.10a). Frechet group reported the synthesis of poly(benzyl ether) dendrimers [26]. Dendrons are synthesized by the step-wise reaction of alkylation and halogenation from the terminal to the core, which is a convergent method (Fig. 3.10b). So far, various kinds of dendrimers such as polyamide, poly(ethyl ether), and polyhydrocarbons dendrimers were reported. PAMAM are commercially available, and various dendrimers with terminal modification were also reported.

Since the dendrimers have specific structures, the physical–chemical properties are totally different from the coiled polymers. The number and density of the terminal are drastically increased with the generation based on the molecular structure. For example, while the intrinsic viscosities of linear polymers increase with molecular weight, those of dendrimer don't increase linearly and are lower than those of linear polymers [27]. The low viscosities implies that dendrimer at the higher generation is less entangled due to the dense packing of the molecular terminals. The viscosities of hyperbranched polymers are also lower than linear polymers, but the relationship with molecular weight is different from that in dendrimers due to the incomplete branching. Other physical properties like refractive indexes also show the unique properties to the molecular weight. Since the dendrimers have an anisotropic structure from core to periphery, the apparent physical properties often depend on the molecular terminal. The solubility of the dendrimers depend on the terminal group. That is, insoluble dendrimers can be dissolved in the solvent by the surface modifi-

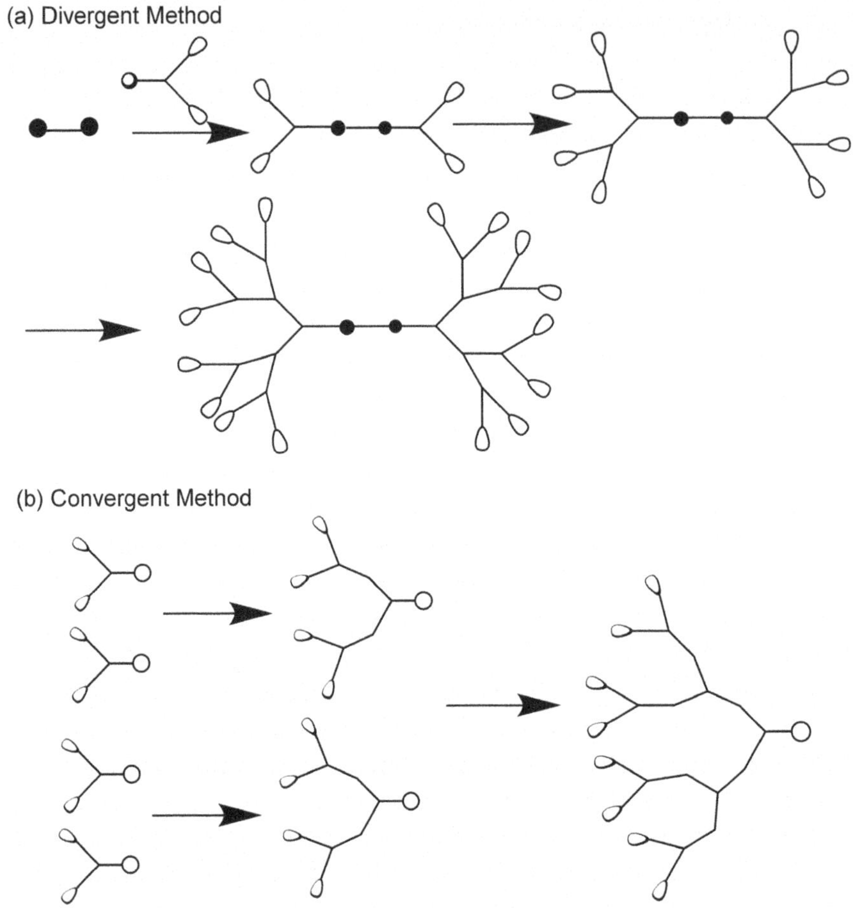

Fig. 3.9 Syntheses of dendrimers by **a** convergent and **b** divergent methods

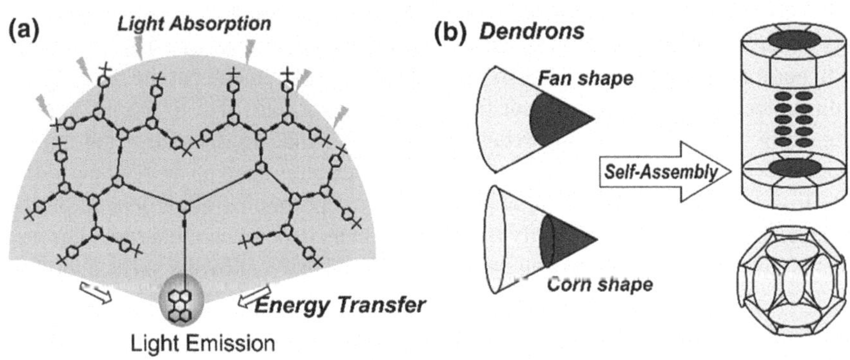

Fig. 3.10 Functional materials with dendrimers: **a** an example of light harvesting dendrimer having antenna effect and **b** the schematic illustration of supramolecules with dendrimer

cation. Poly(benzyl ether) dendrimers with PEG modification are soluble in aqueous solution, and PAMAM dendrimers with alkyl chain terminal are soluble in organic solution [28].

3.1.6.2 Functional Materials with Dendrimers

The fascination of dendrimers is essentially due to the unique architecture and the ordered nanostructure for designing well-defined functional materials. Crooks et al. reported the encapsulation of metals into the dendrimers as hosts [29]. Dendrimers stably adsorbed metals, and dissolve the metals in the various solvents of water, organic solvent, and supercritical CO_2, etc. The metal immobilized on the dendrimers showed high catalytic activities.

Dendrimers have multibranched structure from inward to outward, and the structure are regarded as antenna structure. Some dendrimes exhibited unique photophysical properties like light-harvesting effects [30]. Frechet group reported the efficient luminescence dendrimers (Fig. 3.10a). They prepared polyacetylene and poly(benzyl ether) dendrimers, which are composed of light-collecting chromophores (naphthalene, coumarin 2, etc.) at the periphery and light-emitting molecules (coumarin 343, polythiophene, etc.) at the core. The light energy collected at the periphery of the dendrimers, and the photoinduced energy transfer or energy transfer occurs through the dendrons. The light energy is transferred to the core of dendrimers and that is the light energy amplification due to the antenna like molecular structures.

Dendrimers have also suitable properties for the building blocks of supramolecules. Percec group reported the supramolecular formations with dendrimers, where the dendrimers are modified with long alkyl chain to induce the phase transition and self-assembly (Fig. 3.10b) [31]. They reported the self-assembled columnar structure with liquid crystalline properties of the dendrimer. They attached chromophores to the core of dendrimers and controlled the assemblies, which showed the electrically conducting organic materials. Kato and coworkers reported the rosette structure of columnar dendrimer assemblies by hydrogen bonding, which showed ion channel properties [32].

Dendrimers are also used for biomaterials based on the branched terminal structure and nano size. Since the dendrimers are regarded as small capsules, the drug delivery with dendrimers has been investigated. For example, the dendrimer conjugates with an anticancer drug of doxorubicin showed the efficient cellular uptake to tumor cells, because of the small nano size of the dendrimer capsules. In addition, the dendrimer terminal modifications amplify the target affinity by the effect of cluster ligand. Roy group reported the dendrimers with sialic acid terminals, which inhibited viral adhesion [33]. Kono et al. reported the peptide-modified dendrimers that showed thermo-responsive drug release [34].

3.1.7 Hyperbranched Polymers

Hyperbranched polymers are another type of branched polymer besides dendrimers (Fig. 3.11a). While dendrimers have a uniform structure, hyperbranched polymers have polydispersities that are the mixture of the linear and branched unit due to the insufficient branching. The physical properties of hyperbranched polymers are similar to dendrimers in some extent. Viscosities of linear polymers are known to relate to the molecular weight and become higher in the high molecular weight due to the entanglement of the polymer chain. The viscosities of branched polymers like dendrimers and hyperbranched polymers are much lower than the linear polymers due to the less entanglement. Especially, the viscosities of dendrimers in high generations are very low due to the stiff structure and high terminal density, but those of hyperbranched polymers are modest and increased with molecular weight because terminal densities of hyperbranched polymers are not high due to insufficient branching. The viscosities of hyperbranched polymers are affected by the molecular entanglement in the high molecular weight (Fig. 3.11b).

Though the syntheses of dendrimers are conducted by tedious step-wise reactions, hyperbranched polymers are facilely synthesized by one- or few step reactions, which is advantageous to the commercial use. Hyperbranched polymers have already been put to practical use in industries. Kim and Webstar first reported the hyperbranched polymer in 1990 with AB2 type monomer, [35] using the Suzuki-coupling with transition metal catalyst. Hawker and Frechet et al. reported the hyperbranched polyesters and the hyperbranched polystyrene. Kakimoto et al. reported the syntheses of various polyamide hyperbranched polymers. The hyperbranched polymers are also synthesized by living radical polymerization by the facile one-step reaction. The monomers of hyperbranched polymers have multi, least three, functional groups (ABx).

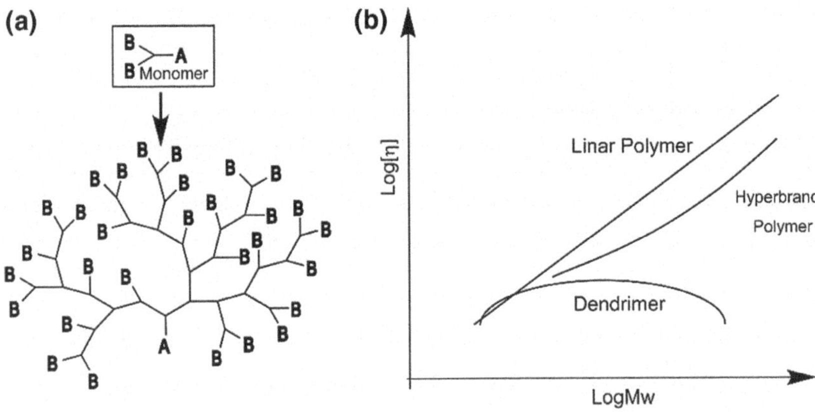

Fig. 3.11 **a** Structure of hyperbranched polymer with linear and branched units. **b** Viscosities of various polymers. Viscosities of hyperbranched polymers were lower than linear polymer and higher than dendrimers

Hyperbranched polymers are much cheaper than dendrimers, and the hyperbranched polymers are put to practical use based on the unique properties. It has been reported the hyperbranched polymers are used as blends composites, electron doping and surface coating due to the low viscosities. The hyperbranched polymers are also used for biomaterials because of multiple terminals of branched polymers. The surface-modified hyperbranched polymers such as ligands and PEG were reported to show efficient molecular recognition and biocompatibility. The hyperbranched polymers are investigated for the various applications such as drug delivery systems and bioimaging.

3.1.8 Polyrotaxane

Polymers with special structures show unique properties. Dendrimers and hyperbranched polymers are also polymers with special structures, but those polymers are based on the covalent bond. On the other hand, polyrotaxanes and polycatenanes have special interlocked structure. Especially, polyrotaxanes are studied by many groups.

Polyrotaxanes are mechanically interlocked molecules consisting of linear molecules and macrocyclic molecules. Apart from usual linear polymers, the inside linear molecules thread macrocycles as necklace structure, where macrocycles have mobility along the inside linear polymers. The unique necklace like structure provides the fabrication of the molecular machines and the actuators (Fig. 3.12). Polyroxanes had been studied from the 1960s, but in the early studies, the yield of rotaxane was very low. Harada et al. prepared polyrotaxane base on the host–guest chemistry, and utilized poly(ethylene glycol) as linear polymer and cyclodextrin as a macrocycle [36]. Stoddart is investigating the molecular system with polyrotaxane, where macrocycles are threaded along the linear polymer based on the redox-reduction response [37]. They controlled the threading of macrocycles by redox-reduction systems, which are applied to the molecular devices such as molecular switch, molecular motors. Ito et al. reported the gels with polyrotaxanes, where the threading macrocycles were utilized as a crosslinker. The polyrotaxane gels showed the stretching properties based on the sliding crosslinkers [38]. Yui group reported biomaterials with polyrotaxanes, where the dynamic properties of polyrotaxanes have advantageous to the molecular recognition and cell cultivations [39].

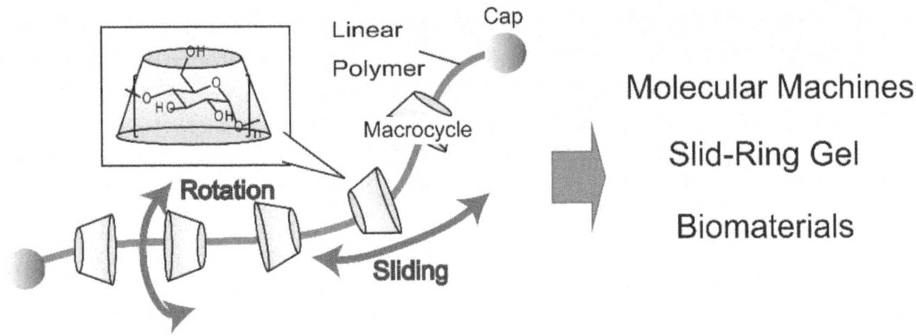

Fig. 3.12 Schematic illustration of polyrotaxane and materials with polyrotaxanes

3.2 Nano- and Microfabrication of Soft Interface

3.2.1 Intelligent Soft Interfaces for Biology

In the medical field, microdevices possessing polymer and protein-modified surface allow point-of-care diagnosis, with a high degree of accuracy at the patient's bedside, the physician is able to make better patient management decisions. Microfabrication techniques were also developed for applications in biology especially in vitro cell and tissue engineering. Cell functions can be modulated by an intricate architecture of cells, chemical properties of the surface, and interaction with modified biomolecules on a micrometer scale. Until now, in vitro cellular interactions were mainly studied by random seeding over homogeneous substrates. With the incorporation of microfabrication technology into biology, it is now possible to design intelligent soft interfaces that reproduce some of the aspects of cellar architecture in vivo.

Cell adhesion to synthetic materials is critically important in various aspects of the development of biomedical devices, artificial organs, and biosensors. Cell adhesion behaviors onto surfaces have been determined for various surface properties including wettability, roughness, surface charge, hydrophobicity, and chemical functionality. When the substrate is exposed to cell suspension in a culture medium containing serum, serum proteins rapidly adsorb onto the surface, and then cells subsequently approach and settle on the surface. Cells adhere to the adsorbed protein layer on the surface. Although it has been accepted that surface properties of materials affect cellular behavior through the adsorbed protein layer, there are yet many uncertain aspects to be resolved for detailed processes of cell adhesion to material surfaces and its correlation to surface properties.

Fabrication methods for constructing intelligent interfaces for cell culture can be divided roughly into two groups: *top-down* and *bottom-up* methods. *Top-down* methods start with patterns made on a large scale and reduce its lateral dimensions before forming micro- and nanostructures. By contrast, *bottom-up* methods begin with atoms or molecules to build up micro- and nanostructures through the use of self-organization described in the early chapter. This paragraph focuses on the typical micro- and nanofabrication techniques in the biological field related to cell patterning.

3.2.1.1 Cell Culturing on Surfaces Constructed by *Top-Down* Method

The important *top-down* techniques for constructing micro- and nanostructures are photolithography, soft lithography, film deposition, etching, and bonding. Photolithography is commonly used to transfer a user-generated shape onto a material surface through the selective exposure of a photoreactive polymer or photopolymerization of photoreactive monomer using UV (Fig. 3.13). Soft lithography encompasses three different techniques, which are all based on the generation and utilization of the mold of a microstructure out of poly(dimethyl siloxane) (PDMS). Film deposition consists of the formation of micron-thick films on the surface of a substrate. Etching selectively removes materials from the surface of the microdevice by either chemical or physical processes. Finally, bonding adheres substrates together with or without the use of intermediary layers.

Cell micropatterning is a method for controlling the placement of living cells on a substrate surface [40]. For instance, since primary hepatocytes (liver cells) play many important roles in various metabolic pathways in vivo, the possibility of using chips covered with hepatocyte arrays in cell-based assay systems as drug screening tools have been investigated [41]. However, primary hepatocytes are well known to lose much of their hepatic functions within the first 2 days of monolayer culturing [42]. Thus, the most crucial issues in the cell or tissue culturing in vitro are long-term viability, the upregulation, and retention of cell functions in vitro on the supporting surface [43].

It is a well known fact that multicellular spheroids exhibit a characteristic in vivo-like morphology; this is attributed to the retention of the 3D architecture and establishment of important cell–cell contacts. Actually, the spheroid patterned array

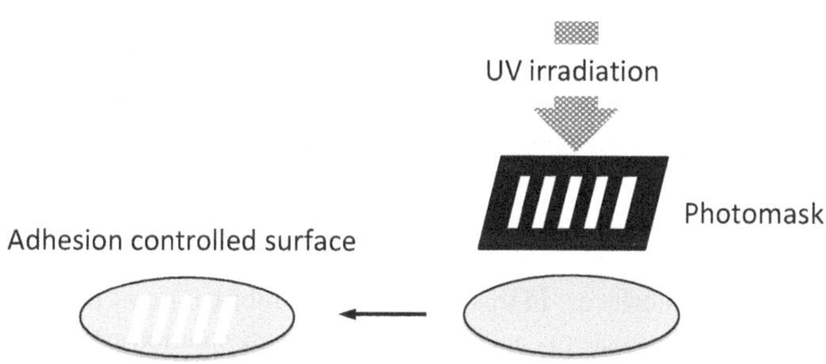

Fig. 3.13 Schematic illustration for constructing adhesion-controlled surface using photolithography

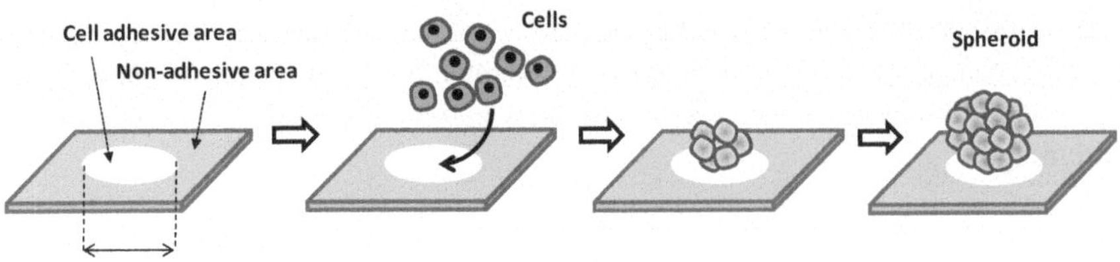

Fig. 3.14 Schematic illustration of multicellular spheroid formation on adhesion-controlled surface

culture of rat primary hepatic cells retains cellular activity for more than 1 month, if bovine aorta endothelial cells (BAECs) are used as feeder cell in 100-mm patterned domains [44]. Figure 3.14 shows the illustration of spheroid formation on adhesion-controlled surface, which has adhesive area and non-adhesive area with micrometer diameter. The synthetic polymers that have non-folding characters are modified on the adhesive surface (e.g., glass surface and polystyrene surface) through covalent and non-covalent bonds, resulting for constructing non-adhesive area.

Among the hepatic cells, fetal mouse liver cells (FMLCs) have been studied as a new material. Although some researchers have surveyed the hepatic activity of FMLCs in monolayer cultures or 3D cultures, such as gel encapsulation cultures, porous reticulated polyvinyl formal resin, and cultures in the spheroid formation of FMLCs. The culturing of FMLC spheroids in a two-dimensional array have reported on a poly(ethylene glycol) (PEG) gel micropatterned surface [45] and evaluate the activity of the FMLCs and the efficiency of the differentiation induction. Interestingly, the FMLCs spheroid upregulated the hepatic activity and differentiation efficiency was improved. Mesenchymal stem cell (MSC) spheroid have also reported and investigated the changes in the secretion level of wound-healing-related proteins [46]. Compared with ADSC cultured monolayer, the spheroid formation on micropatterned surface upregulated the secretion level of these proteins. In the case of MSC, effective reduction of differentiation induction time for osteoblast cells was accomplished using a microdomain patterned surface, where cell accumulations accelerate the osteogenic differentiation in gene expression level [47]. These results indicate the importance of three-dimensional cell architectures constructed on the two-dimensional micropatterned surfaces.

3.2.1.2 Cell Culturing on Surfaces Constructed by *Bottom-Up* Method

The *bottom-up* approaches using low-molecular-weight compounds, biomacromolecules, and materials based on physical/chemical growth through their assembly and adsorption onto interfaces. In this approach, the fabrication of homogeneous or heterogeneous cell aggregates is accomplished through solution processing, which includes superstructure formation with linking through macromolecules, biomacromolecules, and nanomaterials. The fabrication processes in *bottom-up* method mostly depend on the chemical and physical reaction in solution, thus the fabrication of soft

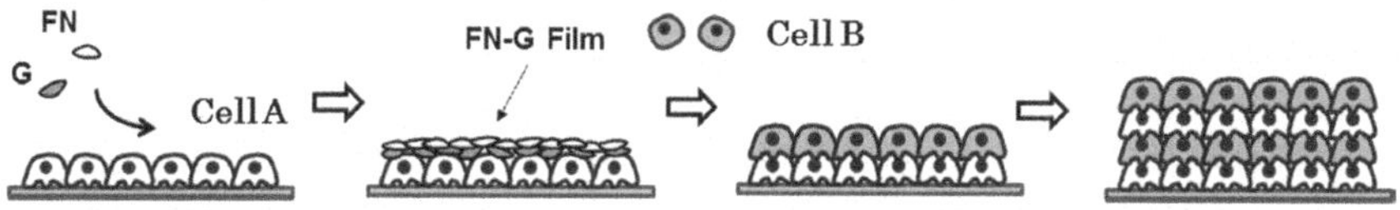

Fig. 3.15 Schematic illustration for constructing three-dimensional multilayered cell architecture using layer-by-layer assembly

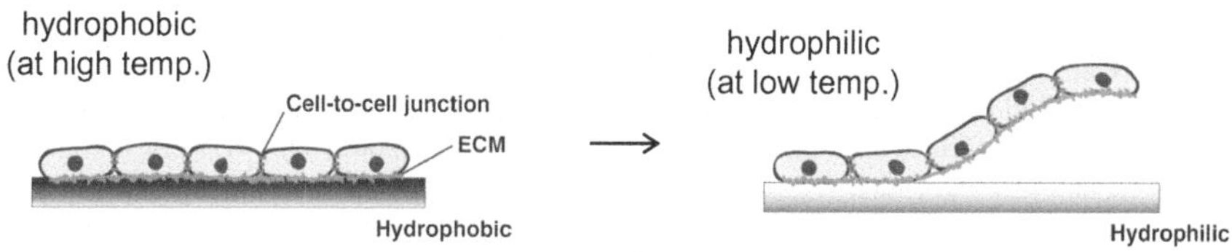

Fig. 3.16 Schematic illustration cell sheet detachment on thermo-responsive PIPAAm-modified surface

interface is accomplished by a simple operation without special equipment. For example, Matsusaki and co-workers established three-dimensional multilayer of cells as shown in Fig. 3.15. They used this technique to construct vessel-like endothelial cell tube in vitro [48]. Liu et al. accomplished cell–cell contact using multivalent bispecific DNA aptamers. DNA aptamer is an oligonucleotide that has a great affinity to target molecules. In their study, bi-, tri-, and tetravalent linkers were used as a scaffold for conjugating multivalent DNA aptamers, where jurkat and ramos cells were contacted effectively [49]. Souza et al. proposed a three-dimensional cell aggregation technique by magnetic cell levitation using nanoparticle-containing hydrogel [50]. The magnetic iron oxide particles were incorporated into hydrogel and cell culture demonstrated under a magnet. Cells were aggregated spherically and forms toroid structure depending on magnet figures. Okano and co-workers developed a thermo-responsive polymer-modified surface for cell sheet engineering, which shows drastic hydrophobicity changes in aqueous solutions [51]. poly(*N*-isopropyl acrylamide) (PIPAAm) is well known as thermo-responsive synthetic polymer and cell attachment property of PIPAAm-modified surface can be controlled by temperature alternation. When cells are seeded on PIPAAm-modified surface at normal culture temperature 37 °C, cells attach and adsorb onto the surface through cell–ECM adhesion, where fiber proteins are adsorbed onto hydrophobic PIPAAm-modified surface. The change to a temperature lower than critical point induces cell detachment as a cell sheet as shown in Fig. 3.16. In these studies, chemical technologies based on *bottom-up* approaches using macromolecules and biomacromolecules play an important role in tissue engineering.

References

1. (a) Seeman N (1982) Nucleic acid junctions and lattices. J Theoretical Boil 99:237–247. (b) Winfree E, Winfree E, Liu F, Wenzler L, Seeman NC (1998) Design and self-assembly of two-dimensional DNA crystals. Nature 394:539–544
2. Rothemund P (2006) Folding DNA to create nanoscale shapes and patterns. Nature 440:297–302
3. Branden C, Tooze J (1999) Introduction to protein structure, Garland Science
4. Scott R, Degrado W, Tew G (2008) De novo designed synthetic mimics of antimicrobial peptides. Curr Opin Biotechnol 19:620–627
5. Aggeli A, Bell M, Boden N, Keen J, Knowles P, McLeish T, Pitkeathy M, Radford S (1997) Responsive gels formed by the spontaneous self-assembly of peptides into polymeric beta-sheet tapes. Nature 386:259–262
6. Nowak AP, Breedveld V, Pakstis L, Ozbas B, Pine DJ, Pochan D, Deming TJ (2002) Rapidly recovering hydrogel scaffolds from self-assembling diblock copolypeptide amphiphiles. Nature 417:424–428
7. Lehn J-M (1988) Supramolecular chemistry Scope and perspectives molecules, supermolecules, and molecular devices (Nobel Lecture). Angew Chem Int Ed 27:89–112
8. Zerkowski JA, MacDonald JC, Seto CT, Wierda DA, Whitesides GM (1994) Design of organic structures in the solid state: molecular tapes based on the network of hydrogen bonds present in the cyanuric acid. cntdot. melamine complex. J Am Chem Soc 116:2382–2391
9. Bohanon TM, Denzinger S, Fink R, Paulus W, Ringsdorf H, Weck M (1995) Barbituric Acid/2,4,6-Triaminopyrimidine aggregates in water and their competitive interaction with a monolayer of barbituric acid lipids at the gas-water interface. Angew Chem Int Ed 34:58–60
10. Hirschberg J, Brunsveld L, Ramzi A, Vekemans J, Sijbesma R, Meijer E (2000) Helical self-assembled polymers from cooperative stacking of hydrogen-bonded pairs? Nature 407:167–170
11. Hasenknopf B, Lehn J-M, Kneisel BO, Baum G, Fenske D (1996) Self-assembly of a circular double helicate. Angew Chem Int Ed 35:1838–1840
12. Fujita M, Oguro D, Miyazawa M, Oka H, Yamaguchi K, Ogura K, (1995) Self-assembly of ten molecules into nanometre-sized organic host frameworks? Nature 378:469–471
13. Batten SR, Champness NR, Chen X-M, Garcia-Martinez J, Kitagawa S, Öhrström L, O fKeeffe M, Suh MP, Reedijk J (2013) Terminology of Metal-Organic Frameworks and Coordination Polymers (IUPAC Recommendations 2013). Pure Appl Chem 85:1715–1724
14. Antonietti M, Förster S, (2003)Vesicles and liposomes: a self-assembly principle beyond lipids. Adv Mater 15:1323–1333
15. Kunitake T (1992) Synthetic bilayer membranes: molecular design, self-organization, and application. Angew Chem Int Ed 31:709–726
16. Bae Y, Fukushima S, Harada A, Kataoka K (2003) Design of environment-sensitive supramolecular assemblies for intracellular drug delivery: polymeric micelles that are responsive to intracellular pH change. Angew Chem Int ed. Ed 42:4640–4643
17. Kimizuka N (2000) Towards self-assembling inorganic molecular wires. Adv Mat 12:1461–1463
18. Miyashita T (1993) Recent studies on functional ultrathin polymer films prepared by the langmuir-blodgett technique. Prog Polym Sci 18:263–294
19. Ulman A (1996) Formation and structure of self-assembled monolayers. Chem Rev 96:1533–1554
20. Netzer L, Ssagiv J (1983) A new approach to construction of artificial monolayer assemblies 105:674–676
21. Decher G (1997) Fuzzy nanoassemblies: toward layered polymeric multicomposites. Science 277:1232–1237
22. Lvov Y, Ariga K, Ichinose I, Kunitake T (1995) Assembly of multicomponent protein films by means of electrostatic layer-by-layer adsorption. J Am Chem Soc 117:6117–6123
23. Kim BS, Park S, Hammond P (2008) Hydrogen-bonding layer-by-layer-assembled biodegradable polymeric micelles as drug delivery vehicles from surfaces. ACS Nano 2:386–392

24. Serizawa T, Hamada K, Kitayama T, Fujimoto N, Hatada K, Akashi M (2000) Stepwise stereocomplex assembly of stereoregular poly(methyl Methacrylate)s on a substrate. J Am Chem Soc 122:1891–1899
25. Tomalia DA, Baker H, Dewald J, Hall M, Kallos G, Martin S, Roeck J, Ryder J, Smith PA (1985) New class of polymers: starburst-dendritic macromolecules. Polym J 17:117–132
26. Hawker CJ, Frechet JMJ (1990) Preparation of polymers with controlled molecular architecture. A new convergent approach to dendritic macromolecules. J Am Chem Soc 112:7638–7647
27. Mourey TH, Turner SR, Rubinstein M, Frechet JMJ, Hawker CJ, Wooley KL (1992) Unique behavior of dendritic macromolecules: intrinsic viscosity of polyether dendrimers. Macromolecules 25:2401–2406
28. Liu M, Kono K, Fréchet J (2000) Water-soluble dendritic unimolecular micelles: their potential as drug delivery agents. J Control Release 65:121–131
29. Zhao M, Crooks RM (1999) Homogeneous hydrogenation catalysis with monodisperse, dendrimer-encapsulated Pd and Pt nanoparticles. Angew Chem Int Ed 38:364–366
30. Adronov A, Fréchet JMJ (2000) Light-harvesting dendrimers. Chem Commun, 1701–1710
31. Rosen B, Wilson C, Wilson D, Peterca M, Imam M, Percec V (2009) Dendron-mediated self-assembly, disassembly, and self-organization of complex systems. Chem Rev 109:6275–6540
32. Sakai N, Kamikawa Y, Nishii M, Matsuoka T, Kato T, Matile S (2006) Dendritic folate rosettes as ion channels in lipid bilayers. J Am Chem Soc 128:2218–2219
33. Reuter J, Myc A, Hayes M, Gan Z, Roy R, Qin D, Yin R, Piehler L, Esfand R, Tomalia D, Baker JR (1999) Inhibition of viral adhesion and infection by sialic-acid-conjugated dendritic polymers. Bioconjugate Chem 10:271–278
34. Kono K, Murakami E, Hiranaka Y, Yuba E, Kojima C, Harada A, Sakurai K (2011) Thermosensitive molecular assemblies from poly(amidoamine) dendron-based lipids. Angew Chem Int 50:6332–6336
35. Kim YH, Webster OW (1992) Hyperbranched polyphenylenes. Macromolecules 25:5561–5572
36. Harada A, Li J, Kamachi M (1992) The molecular necklace: a rotaxane containing many threaded α-cyclodextrins. Nature 356:325–327
37. Balzani V, Credi A, Raymo FM, Stoddart JF (2000) Artificial molecular machines. Angew Chem Int Ed 39:3348–3391
38. Okumura Y, Ito K (2001) The polyrotaxane gel: a topological gel by figure-of eight cross-links. Adv Mat 13:485–487
39. Seo JH, Kakinoki S, Inoue Y, Nam K, Yamaoka T, Ishihara K, Kishida A, Yui N (2013) The significance of hydrated surface molecular mobility in the control of the morphology of adhering fibroblasts. Biomaterials 4:3206–3214
40. (a) Mrksich M, Whitesides GM (1996) Annu Rev Biophys Biomol Struct 25, 55. (b) Folch A, Toner M (2000) Annu Rev Biomed Eng 2, 227. (c) Jung DR, Kapur R, Adams T, Giuliano KA, Mrksich M, Craighead HG, Taylor DL (2001) Crit Rev Biotechnol 21, 111. (d) Bhadriraju K, Chen CS (2002) Drug Discov Today 7, 612. (e) Mrksich M (2002) Curr. Opin Chem Biol 6, 794. 6. Corey JM, Feldman EL (2003) Exp Neurol 184, S89. (f) Falconnet D, Csucs G, Grandin HM, Textor M (2006) Biomaterials 27, 3044. (g) Yap FL, Zhang Y (2007) Biosens Bioelectron 22, 775
41. Otsuka H, Hirano A, Nagasaki Y, Okano T, Horiike Y, Kataoka K (2004) Chem Bio Chems 5, 850–855. (a) Alvarez, SD, Derfus AM, Schwartz MP, Bhatia SN, Sailor MJ (2009) Biomaterials 30, 26–34; (b) Nelson CM, Inman JL, Bissell MJ (2008) Nature Protocols 3, 674–678
42. Li AP, Beck DJ, Colburn SM (1992) Vitro Cell Dev Biol A 28:673–677
43. LeCluyse E, Bullock P, Parkinson A (1996) Adv Drug Deliv Rev 22:133–186
44. Ikeda Y, Jomura T, Horiuchi U, Saeki J, Yoshimoto K, Ikeya T, Nagasaki Y (2012) Long-term survival and functional maintenance of hepatocytes by using a microfabricated cellarray. Colloids Surf B Biointerfaces 97:97–100
45. Kojima R, Yoshimoto K, Miyoshi H, Nagasaki Y (2009) Spheroid array of fetal mouse liver cells constructed on a PEG-gel micropatterned surface: upregulation of hepatic functions by co-culture with nonparenchymal liver cells. Lab Chip 9:1991–1993
46. Furuhata Y, Kikuchi Y, Tomita S, Yoshimoto K (2016) Genes Cells 21(12):1380–1386

47. Furuhata Y, Yoshitomi T, Kikuchi Y, Sakao M, Yoshimoto K, (2017) ACS Appl Mater Interfaces 9:9339
48. Nishiguchi A, Yoshida H, Matsusaki M, Akashi M (2011) Adv. Mater. 23:3506–3510
49. Liu X, Yan H, Liu Y, Chang Y (2011) Small 7(12):1673–1682
50. Souza GR, Molina JR, Raphael RM, Ozawa MG, Stark DJ, Levin CS, Bronk LF, Ananta JS, Mandelin J, Georgescu M-M, Bankson JA, Gelovani JG, Killian TC, Arap W, Pasqualini R (2010) Nat Nanotechnol 5:291–296
51. Matsuda N, Shimizu T, Yamato M, Okano T (2007) Adv Mater 19:3089–3099

Part III
Characterization and Physical Properties of Soft Interface

Chapter 4
Infrared and Raman Spectroscopy for Thin-Film Analysis

Takeshi Hasegawa

4.1 Infrared Spectroscopy

According to quantum mechanics, the light absorption by a dipole oscillation is theorized by Fermi's golden rule [1]. The most important part of the rule is the transition integral, $\langle k|\hat{H}'|j\rangle$, where k and j correspond to the final and the initial states of an energy transition, and $\hat{H}'$ the perturbation Hamiltonian. The perturbation factor can be expressed in a good approximation manner by

$$\hat{H}' = \mathbf{p} \cdot \mathbf{E} \tag{4.1}$$

where the vectors, **p** and **E**, correspond to the dipole moment and the electric field at the dipole, respectively. In general, the wavelength of the light is much longer than the dipole size, the transition integral can be rewritten as

$$\langle k|\mathbf{p} \cdot \mathbf{E}|j\rangle = \langle k|\mathbf{p}|j\rangle \cdot \mathbf{E} \tag{4.2}$$

As a result, the band intensity is proportional to the dot product of the transition moment, $\langle k|\mathbf{p}|j\rangle$, and the electric field. Of note here is that electric field is not of the irradiating light, but the electric field 'at the dipole.' The electric field in a thin film is largely influenced by the reflection at an interface, which cannot be monitored by a spectrometer. As a result, the band intensity remains largely influenced by the substrate surface even after making a ratio of the sample and the background (substrate alone) spectra.

The correlation between the band intensity of IR absorption and the tilt angle of the transition moment is fully theorized by calculating the electric field distribution near the interface, but for an intuitive understanding, the surface selection rule based on the

T. Hasegawa (✉)
Institute for Chemical Research, Kyoto University, Uji, Kyoto, Japan
e-mail: htakeshi@scl.kyoto-u.ac.jp

M. Maeda et al. (eds.), *Molecular Soft-Interface Science*,
https://doi.org/10.1007/978-4-431-56877-3_4

electric field analysis is much more convenient. Deduction of a mathematical formula of absorption spectra of a thin film deposited on a substrate is a difficult task [1, 2], since the analytical calculation is done by employing Abele's transfer matrix method, which makes the equations highly complicated. Fortunately, however, representative optical configurations have already been analyzed to have useful formulae [1, 3].

The most fundamental cases are the normal incidence transmission measurements on a transparent material, and the reflection measurements on a metallic surface, which are called normal incidence transmission (Tr) and reflection–absorption (RA) spectrometries, respectively.

$$A_{\mathrm{Tr}} = \frac{1}{l_n 10.\lambda} \frac{8\pi d}{n_1 + n_3} \mathrm{I_m}(\varepsilon_{\mathrm{r}}, x) \tag{4.3}$$

$$A_{\mathrm{RA}}^{\mathrm{p}} = \frac{1}{l_n 10.\lambda} n_1^3 \frac{sin^2\theta}{cos\theta} \mathrm{I_m}\left(-\frac{1}{\varepsilon_{\mathrm{r}}, z}\right) \tag{4.4}$$

Here, ε_{r} and d are electric relative permittivity and the thickness of the thin film, respectively, and θ is the angle of incidence measured from the surface normal. The superscript, p, on A_{RA} represents the p-polarization, whose electric field vector is involved in the incidental plane spanned by the surface normal and the incidence vectors. These equations are good approximations for a very thin film ($d/\lambda = 1$). As found in the equations, the shapes of the absorbance spectra of a structurally isotropic sample, A_{Tr} and $A_{\mathrm{RA}}^{\mathrm{p}}$w, depend on the two functions, $\mathrm{Im}(\varepsilon_{\mathrm{r}}, x)$ and $\mathrm{Im}(-1/\varepsilon_{\mathrm{r}}, z)$, which are called TO and LO energy loss functions, respectively. Since the TO and LO functions correspond to the energy losses caused by molecular vibrations parallel and perpendicular to the film surface, respectively, the Tr and RA spectrometries are concluded as techniques to draw molecular orientation in a thin film. The conclusions are conveniently called 'surface selection rules' of Tr and RA spectrometries [1, 3]. Of note is that the RA measurements are performed on a 'metallic' surface using the p-polarized light. When the substrate is replaced by a dielectric material, the equation should also be replaced by a largely different equation [1, 3].

The permittivity, $\varepsilon(\omega)$, of a dielectric material, which can easily be deduced from a simple consideration of a restricted electron based on the classical dynamics, is formulated by Lorentz's function:

$$\varepsilon_r(\omega) = \varepsilon_{\mathrm{r},\infty} + \frac{\mathrm{e}^2 N_0}{m^*} \sum_j \frac{f_j}{\omega_j^2 - \omega^2 - \mathrm{i}\gamma_j\omega}. \tag{4.5}$$

Here, $\varepsilon_{\mathrm{r},\infty}$ is the relative permittivity at a high-frequency limit, ω_j, γ_j, and f_j are the angular frequency, the damping factor, and the oscillator strength of the jth band, respectively. It is no problem to consider the term of $\mathrm{e}^2 N_0/m^*$ to be involved in the oscillator strength.

With the Lorentz function, the real and imaginary parts of the permittivity having a single band ($j = 1$ only) are calculated as shown in Fig. 4.1a.

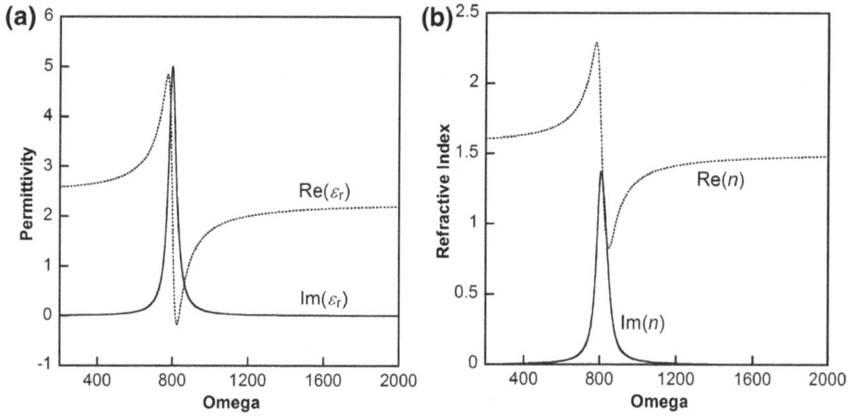

Fig. 4.1 Simulated **a** complex electric permittivity and **b** refractive index using the parameters: $\varepsilon_{r,\infty} = 2.25$, $\omega_1 = 800$, $\gamma_1 = 50$, $f_1 = 2 \times 10^5$

At $\omega_1 = 800$, a single band develops as $\mathrm{Im}(\varepsilon_r)$, whereas the $\mathrm{Re}(\varepsilon_r)$ function appears as a dispersed curve about the band. The degree of dispersion depends on the peak intensity of $\mathrm{Im}(\varepsilon_r)$. Since the permittivity is related to the refractive index, n, by $n^2 = \varepsilon_r$, the complex refractive index is also obtained as in Fig. 4.1b.

When n is denoted as $n \equiv n' + in''$, then $\mathrm{Im}(\varepsilon) = 2n'n''$ holds. By referring to Eqs. 4.3 and 4.4, therefore, the Tr and RA spectra of a thin film prove to be directly influenced by both curves of n' and n''. This is different from the absorption in a bulk sample, in which only n'' attributes the absorption. In this manner, the absorption spectrum of a thin film has a *distorted* band due to the dispersion of n' especially for a strongly absorbing (large n'') band. In practice, the distortion often looks like a 'shift.' The calculated TO and LO functions using the permittivity in Fig. 4.1a are shown in Fig. 4.2.

As found in the figure, the TO and LO peaks appear at different positions, although the band shape looks similar to each other.

We have to note that the band shift comes even from the isotropic layer, i.e., both TO and LO functions are calculated by using a common permittivity. In other words, the band shift happens because the thin film is deposited on an interface, and measured by different optical configurations such as Tr and RA. Therefore, if we find a band shift for two spectra measured on different substrates, we cannot directly attribute the shift to the chemical interaction with (or molecular orientation on) the different substrates. Before making such an inappropriate chemical discussion, we have to remember that the band shift can be generated by the pure electromagnetic (or optical) effect.

As a practical example, IR Tr and RA spectra of Langmuir–Blodgett films of cadmium stearate deposited on a germanium substrate and a gold-evaporated glass, respectively, are presented in Fig. 4.3 [4].

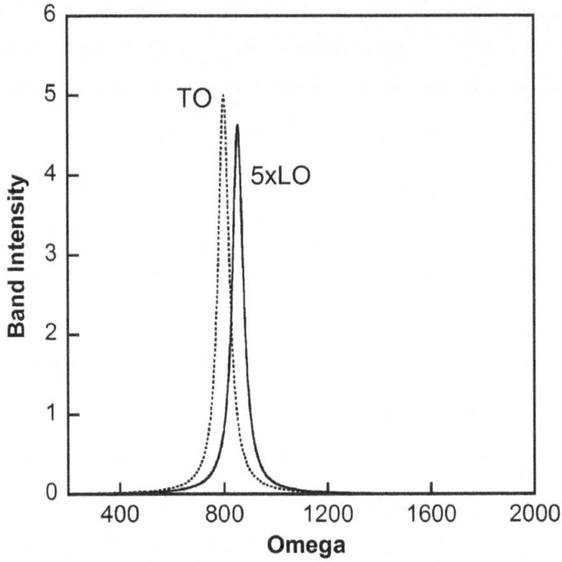

Fig. 4.2 Simulated TO and LO energy loss functions calculated by using the complex permittivity in Fig. 4.1. For better visibility, the LO function is five times magnified

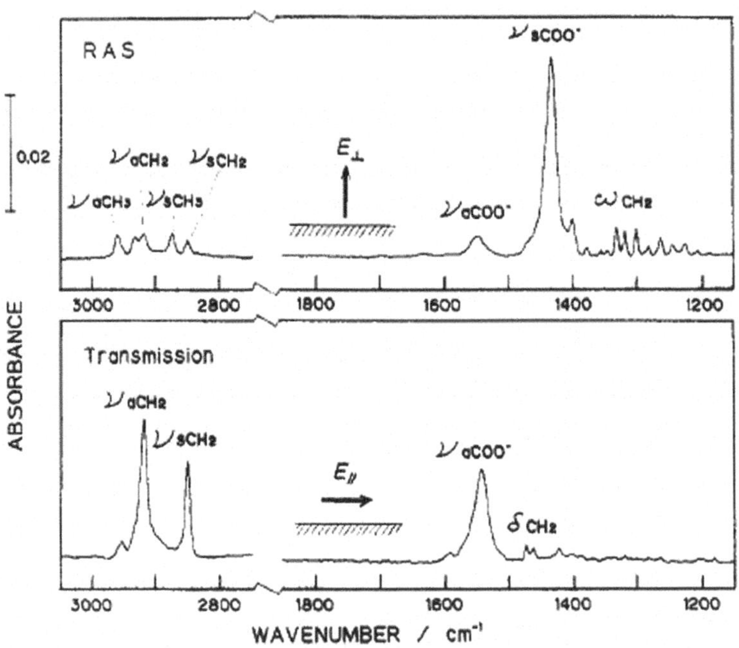

Fig. 4.3 IR RA (top) and Tr (bottom) spectra of seven-monolayer LB film of cadmium stearate on a ZnSe substrate and a silver-evaporated glass slide, indicating the LO and TO functions, respectively. For the RA measurement, the angle of incidence was 85° from the surface normal [4]

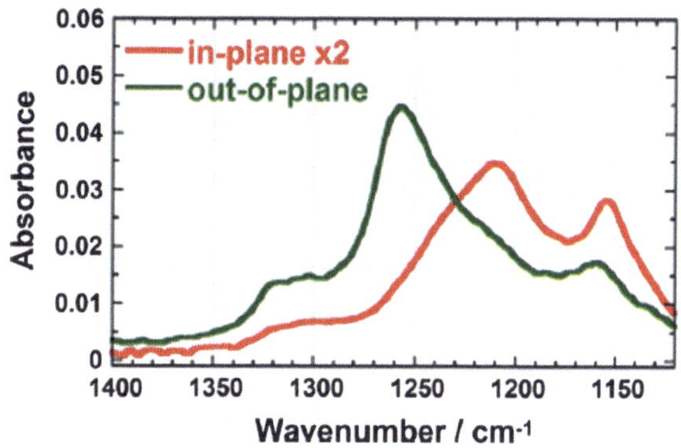

Fig. 4.4 IR MAIRS spectra of a thin (ca. 150 nm) Nafion film prepared on silicon

In this case, the TO and LO functions correspond to the surface selection rules of the Tr and RA techniques, respectively, because of the anisotropic permittivity. The intensities in the two spectra of a band are complementary to each other, which indicates that the molecules are highly oriented. For example, the anti-symmetric and symmetric CH_2 stretching vibration (ν_aCH_2 and ν_sCH_2) modes appeared at 2917 and 2850 cm^{-1} have nearly parallel transition moments to the surface. Since these band locations indicate that the hydrocarbon chain is in an ordered manner with the all-trans zigzag conformation, the molecular axis and the two modes are mutually orthogonal, which further tells us that the molecule stands nearly perpendicular to the surface. In this fashion, the molecular orientation analysis is easily performed when both TO and LO functions are available.

The combination technique of Tr and RA thus works powerfully when an 'identical' film can be prepared on the different substrates. In practice, however, the film deposition is often influenced by the surface property of the substrate. To get over the experimental difficulty, the multiple-angle incidence resolution spectrometry (MAIRS) technique [1, 5, 6] is quite useful.

MAIRS yields a set of two spectra corresponding to the TO and LO functions and at a time, respectively, and the results are obtained from an identical sample. For IR MAIRS measurements, an IR transparent plate with a high refractive index is used as the substrate of a thin film. Germanium (Ge) is the most convenient material for the purpose. If we want to use a substrate with a low refractive index less than 2.5, an alternative technique, pMAIRS [7], is used. For example, a silicon wafer covered by a surface oxide layer is used for the substrate, pMAIRS should be employed, since the silicon oxide has a low refractive index.

Figure 4.4 presents IR MAIRS spectra of a thin Nafion film deposited on a silicon substrate [8], in which the in-plane (IP) and out-of-plane (OP) spectra correspond to the transmission and RA spectra, respectively.

The band at 1260 cm^{-1} that has long been unrecognized appears apparently in the OP spectrum only. After some DFT calculations, this band is assigned to be a vibration mode of the $-SO_3H$ group in a coupled dimer. In this manner, the side chains of Nafion are found to have a dimer structure, which is highly oriented to the surface.

IR spectroscopy thus works quite powerfully to reveal not only the primary structure of the compound, but the molecular interaction and the orientation particularly when an appropriate optical configuration is employed.

4.2 Raman Spectroscopy

Raman spectroscopy is another choice of measuring normal modes in a thin film. The Raman scattering comes from a Raman dipole, **p**, induced by external electric field, **E**, of laser light when the dipole is isolated in space.

$$\mathbf{p}^{\mathrm{L}} = \boldsymbol{\alpha}^{\mathrm{L}}\mathbf{E}^{\mathrm{L}} \tag{4.6}$$

Here, the superscript, L, indicates that the parameters are defined in the laboratory coordinate because the electric field is set on the laboratory coordinate. Since the directions of the dipole and the electric field can be different from each other, the two vectors are correlated with each other by a tensor (matrix in this case), $\boldsymbol{\alpha}^{\mathrm{L}}$, which is called polarizability. The polarizability should be, however, defined on the molecular coordinate to discuss the molecular structure. Therefore, the polarizability on the molecular coordinate, $\boldsymbol{\alpha}^{\mathrm{M}}$, must be correlated with $\boldsymbol{\alpha}^{\mathrm{L}}$.

In a liquid sample, every dipole moves around and $\boldsymbol{\alpha}^{\mathrm{L}}$ can be correlated with a time average of $\boldsymbol{\alpha}^{\mathrm{M}}$. The formulation is very famous [9], which is used to consider the depolarization of Raman scattering, ρ_{p}.

$$\rho_{\mathrm{p}} = \frac{\langle \alpha_{yz}^2 \rangle}{\langle \alpha_{zz}^2 \rangle} = \frac{3\gamma^2}{45\bar{\alpha}^2 + 4\gamma^2} \tag{4.7}$$

Here, the bracket means the average on orientation correlating the molecular coordinate with the laboratory one, which uses the newly introduced two parameters:

$$\bar{\alpha} = \frac{1}{3}\left(\alpha_{xx}^{\mathrm{M}} + \alpha_{yy}^{\mathrm{M}} + \alpha_{zz}^{\mathrm{M}}\right) \tag{4.8}$$

$$\begin{aligned}\gamma^2 = &\frac{1}{2}\Big[\left(\alpha_{xx}^{\mathrm{M}} - \alpha_{yy}^{\mathrm{M}}\right)^2 + \left(\alpha_{yy}^{\mathrm{M}} - \alpha_{zz}^{\mathrm{M}}\right)^2 + \left(\alpha_{zz}^{\mathrm{M}} - \alpha_{xx}^{\mathrm{M}}\right)^2 \\ &+6\left\{\left(\alpha_{xy}^{\mathrm{M}}\right)^2 + \left(\alpha_{yz}^{\mathrm{M}}\right)^2 + \left(\alpha_{zx}^{\mathrm{M}}\right)^2\right\}\Big]\end{aligned} \tag{4.9}$$

On measuring a thin film deposited on a substrate, the conversion of the coordinate becomes more explicit. When the conversion matrix, **U**, is introduced as follows:

$$\mathbf{p}^{\mathrm{L}} = \mathbf{U}\mathbf{p}^{\mathrm{M}} \text{ and } \mathbf{E}^{\mathrm{L}} = \mathbf{U}\mathbf{E}^{\mathrm{M}} \tag{4.10}$$

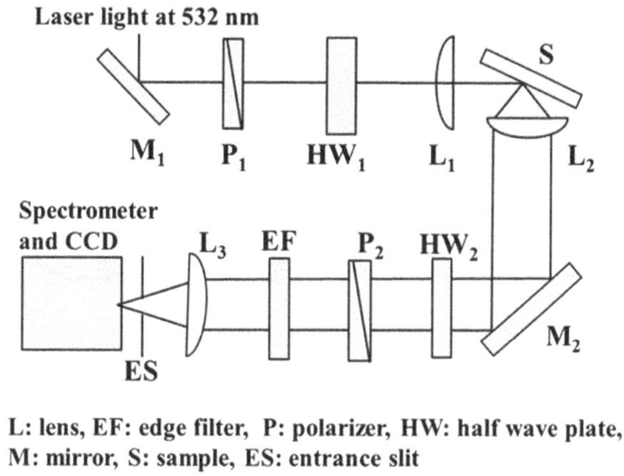

Fig. 4.5 Optical setup for the polarized Raman measurements

Equation 4.6 becomes

$$\mathbf{U}\mathbf{p}^{\mathrm{M}} = \boldsymbol{\alpha}^{\mathrm{L}}\mathbf{U}\mathbf{E}^{\mathrm{M}} \Leftrightarrow \mathbf{p}^{\mathrm{M}} = \mathbf{U}^{-1}\boldsymbol{\alpha}^{\mathrm{L}}\mathbf{U}\mathbf{E}^{\mathrm{M}} \tag{4.11}$$

Therefore, $\boldsymbol{\alpha}^{\mathrm{M}} = \mathbf{U}^{-1}\boldsymbol{\alpha}^{\mathrm{L}}\mathbf{U}$ is obtained. Since the conversion is an expression of the orientation of a dipole observed on the laboratory coordinate, the determination of $\mathbf{U}$ is the determination of molecular orientation. This is the principle of the molecular orientation analysis in a thin film using Raman spectroscopy, which is much more complicated than that using IR spectroscopy.

Here, the determination of $\mathbf{E}^{\mathrm{L}}$ is another complicated task when a thin film on a substrate is analyzed. Since the electric field 'at the dipole' is largely influenced by the interfaces (air/film and film/substrate), the field must be calculated by Abeles' transfer matrix method or its related techniques [2, 3].

The Raman optical scheme for measuring thin films is illustrated in Fig. 4.5. The excitation laser is focused by a lens (L_1) in the sample surface, and the emitted Raman scattering is collected by another lens (L_2). By introducing a half-wave plate, the issue of polarization dependence of the spectrometer is removed [10]. With this system, the depolarization ratio of a non-totally symmetric vibrational mode is very close to 0.75, which guarantees the precision of spectrometer for polarization measurements.

In Fig. 4.6, polarization-dependent non-resonance Raman spectra of a 5-monolayer LB film of cadmium stearate deposited on a glass plate. The LB film was prepared at a surface pressure of 25 mN m^{-1}.

When the two polarizations are orthogonal to each other (SP or PS), the $\nu_s CH_2$ band is largely suppressed because this mode belongs to the totally symmetric ($d^+(0)$) mode. Although the methylene 'group' can be treated as the point group of C_{2v} the methylene 'chain' must be considered on D_{2h}, since Raman detects the coupled vibration with the phase difference of nil. Therefore, in this case, the $\nu_s CH_2$ mode corresponds to A_g. In the same manner, the $\nu_a CH_2$ mode ($d^-(0)$) corresponds to B_{1g}. The polarizability tensors of the two modes can be represented as

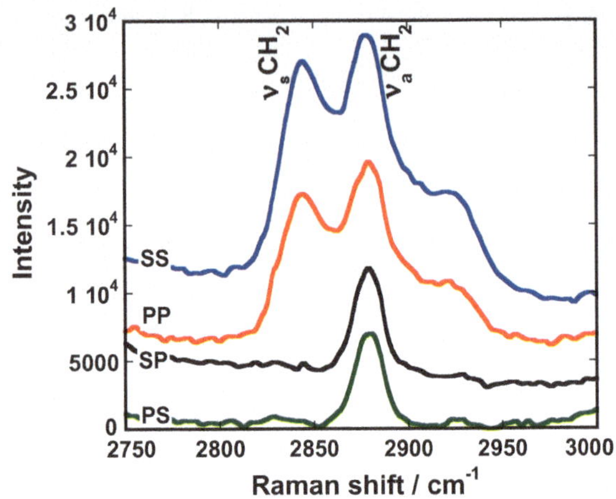

Fig. 4.6 Polarization-dependent Raman spectra of a five-monolayer LB film of cadmium stearate deposited on a glass plate. 'PS' means that the polarization of the incident and collected lights are P and S, respectively

$$\boldsymbol{\alpha}^{\mathrm{M}}[A_g] = \begin{pmatrix} \alpha_{xx}^{\mathrm{M}} & 0 & 0 \\ 0 & \alpha_{yy}^{\mathrm{M}} & 0 \\ 0 & 0 & \alpha_{zz}^{\mathrm{M}} \end{pmatrix} \text{ and } \boldsymbol{\alpha}^{\mathrm{M}}[B_{1g}] = \begin{pmatrix} 0 & \alpha_{xy}^{\mathrm{M}} & 0 \\ \alpha_{yx}^{\mathrm{M}} & 0 & 0 \\ 0 & 0 & 0 \end{pmatrix} \tag{4.12}$$

Since $\alpha_{xy}^{\mathrm{M}} = \alpha_{yx}^{\mathrm{M}} \equiv \alpha$ reasonably holds, the B_{1g} mode is much more convenient for analysis. Therefore, the $\nu_a CH_2$ mode appeared at 2879 cm^{-1} (largely different position from the IR one) is appropriate for quantitative analysis.

According to Snyder [11], when the alkyl chain stands perpendicularly to the surface, the Raman intensity of the SP polarization spectrum is proportional to $\left(\alpha_{xx}^{\mathrm{M}} - \alpha_{yy}^{\mathrm{M}}\right)^2$. Very coincidently, $\alpha_{xx}^{\mathrm{M}} \approx \alpha_{yy}^{\mathrm{M}}$ holds for the alkyl chain, and the complete disappearance is readily explained by the perpendicular orientation of the chain. Note that $\alpha_{xx}^{\mathrm{M}} \approx \alpha_{yy}^{\mathrm{M}}$ does *not* mean the uniaxial orientation ($\alpha_{xx}^{\mathrm{L}} \approx \alpha_{yy}^{\mathrm{L}}$).

For the detail of quantitative analysis of molecular orientation, the reader referred to a reference [10].

References

1. Hasegawa T (2017) Quantitative infrared spectroscopy for understanding of a condensed mater. ET Springer, Tokyo
2. Yeh P (1998) Optical waves in layered media. Wiley, Hoboken, NJ
3. Tolstoy VP, Chernyshova IV, Skryshevsky VA (2003) Wiley. Hoboken, NJ
4. Umemura J, Kamata T, Kawai T, Takenaka T (1990) Quantitative evaluation of molecular orientation in thin Langmuir-Blodgett films by FT-IR transmission and reflection-absorption spectroscopy. J Phys Chem 94:62–67
5. Hasegawa T (2002) A novel measurement technique of pure out-of-plane vibrational modes in thin films on a nonmetallic material with no polarizer. J Phys Chem B 106:4112–4116
6. Hasegawa T (2008) A new approach to analysis of molecular structure in thin films: infrared multiple-angle incidence resolution spectrometry. Appl Spectrosc Rev 43:181–201

7. Hasegawa T (2008) Advanced multiple-angle incidence resolution spectrometry for thin-layer analysis on a low-refractive-index substrate. Anal Chem 79:4385–4389
8. Nagao Y (2013) Highly oriented sulfonic acid groups in a Nafion thin film on si substrate. J Phys Chem C 117:3294–3297
9. Long DA (2002) The raman effect: a unified treatment of the theory of raman scattering by molecules. Wiley, Chichester
10. Itoh Y, Hasegawa T (2012) Polarization dependence of raman scattering from a thin film involving optical anisotropy theorized for molecular orientation analysis. J Phys Chem A 116:5560–5570
11. Snyder RG (1971) Raman scattering activities for partially oriented molecules. J Mol Spectrosc 37:353–365

Chapter 5
Sum Frequency Generation (SFG)

Daisuke Kawaguchi and Keiji Tanaka

5.1 Introduction

Sum frequency generation (SFG) vibrational spectroscopy has been used to study molecular structures at surfaces and interfaces [1–7]. To obtain SFG signals, centrosymmetry of the system must be broken. This condition can be satisfied only at interfaces, leading to an SFG signal that is highly interface specific. The surface and interfacial sensitivities of SFG are much better than those of other techniques such as attenuated total reflectance Fourier-transform infrared spectroscopy [8] and surface-enhanced Raman spectroscopy [9]. The other intriguing advantage of SFG vibrational spectroscopy is the ability to gain deep understanding of the orientation of functional groups by using different polarization combinations of input and output beams. Thus, SFG enables us to study in situ local conformation of polymer chains at various interfaces.

5.2 Principle

The SFG intensity (I^{SFG}) is proportional to the square of the absolute value of the effective sum frequency susceptibility tensor ($\chi^{(2)}_{\mathrm{eff}}$), which is itself related to the second-order nonlinear susceptibility tensor $\chi^{(2)}$ in the laboratory coordinate system, as well as the intensities of the two input beams ($I_{\mathrm{i}}(\omega_{\mathrm{Vis}})$ and $I_{\mathrm{i}}(\omega_{\mathrm{IR}})$). I^{SFG} in the reflective direction can be formulated as [3]

$$I^{SFG} = \frac{8\pi^3\omega^2\sec^2\theta_{SF}}{c_0^3 n_i(\omega_{SF})n_i(\omega_{Vis})n_i(\omega_{IR})}\left|\chi^{(2)}_{eff}\right|^2 I_i(\omega_{Vis})I_i(\omega_{IR})AT \tag{5.1}$$

D. Kawaguchi · K. Tanaka (✉)
Department of Applied Chemistry, Kyushu University, Fukuoka 819-0395, Japan
e-mail: k-tanaka@cstf.kyushu-u.ac.jp

M. Maeda et al. (eds.), *Molecular Soft-Interface Science*,
https://doi.org/10.1007/978-4-431-56877-3_5

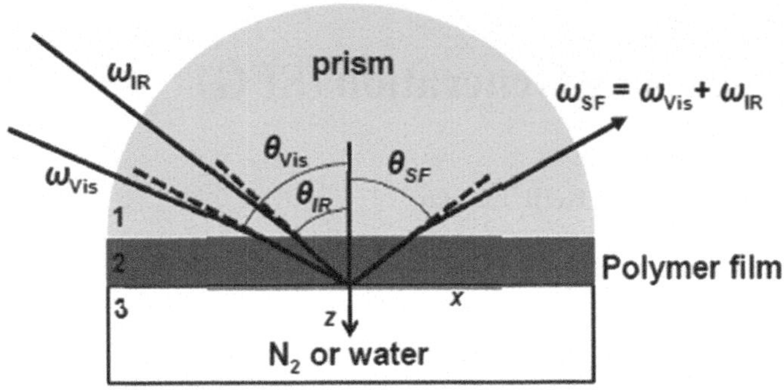

Fig. 5.1 A schematic representation of the optical geometry used in a SFG spectroscopy experiments (Reproduced from Ref. [11] with permission from The Royal Society of Chemistry)

where $n_i(\omega)$ is the refractive index of medium i at frequency ω, θ_{SF} is the reflection angle of the sum frequency field, and c_0 is the speed of light in the vacuum. In Eq. 5.1, A and T are the beam overlap cross-section at the interface and the pulse width, respectively. In the case of an azimuthally isotropic interface, there are only four independent nonvanishing components of $\chi^{(2)}$. With the laboratory coordinates chosen such that z is along the interface normal and x is in the incidence plane, they are $\chi_{xxz} = \chi_{yyz}$, $\chi_{xzx} = \chi_{yzy}$, $\chi_{zxx} = \chi_{zyy}$, and χ_{zzz}. Here, SFG spectra shown were collected with the *ssp* and *ppp* polarization combinations. The $\chi^{(2)}_{\text{eff}}$ for the two polarization combinations can be expressed as follows [3]:

$$\chi^{(2)}_{eff,ssp} = L_{yy}(\omega_{SF})L_{yy}(\omega_{Vis})L_{zz}(\omega_{IR})\sin\theta_{IR}\chi_{yyz} \tag{5.2}$$

$$\begin{aligned}\chi^{(2)}_{eff,ppp} = &-L_{xx}(\omega_{SF})L_{xx}(\omega_{Vis})L_{zz}(\omega_{IR})\cos\theta_{SF}\cos\theta_{Vis}\sin\theta_{IR}\chi_{xxz}\\ &-L_{xx}(\omega_{SF})L_{zz}(\omega_{Vis})L_{xx}(\omega_{IR})\cos\theta_{SF}\sin\theta_{Vis}\cos\theta_{IR}\chi_{xzx}\\ &+L_{zz}(\omega_{SF})L_{xx}(\omega_{Vis})L_{xx}(\omega_{IR})\sin\theta_{SF}\cos\theta_{Vis}\cos\theta_{IR}\chi_{zxx}\\ &+L_{zz}(\omega_{SF})L_{zz}(\omega_{Vis})L_{zz}(\omega_{IR})\sin\theta_{SF}\sin\theta_{Vis}\sin\theta_{IR}\chi_{zzz}\end{aligned} \tag{5.3}$$

where $L_{ii}(i = x, y, z)$ is the Fresnel coefficients. The angle between the surface normal and the beam is expressed as θ, as shown in Fig. 5.1. Subscripts Vis and IR denote the input beams. Here, a polymer thin film prepared on a prism and contacting with N_2 or water is taken as an example. To calculate the Fresnel factor for polymer/N_2 or polymer/water interface, the following equations are used [10],

$$L_{xx}(\omega) = \frac{2n_1(\omega)\cos\theta_1}{n_2(\omega)\cos\theta_1 + n_1(\omega)\cos\theta_2} \times \frac{2n_2(\omega)\cos\theta_2}{n_3(\omega)\cos\theta_2 + n_2(\omega)\cos\theta_3} \times \frac{\cos\theta_3}{\cos\theta_1} \tag{5.4}$$

$$L_{yy}(\omega) = \frac{2n_1(\omega)\cos\theta_1}{n_1(\omega)\cos\theta_1 + n_2(\omega)\cos\theta_2} \times \frac{2n_2(\omega)\cos\theta_2}{n_2(\omega)\cos\theta_2 + n_3(\omega)\cos\theta_3} \tag{5.5}$$

$$L_{zz}(\omega) = \frac{2n_1(\omega)\cos\theta_1}{n_2(\omega)\cos\theta_1 + n_1(\omega)\cos\theta_2} \times \frac{2n_2(\omega)\cos\theta_2}{n_3(\omega)\cos\theta_2 + n_2(\omega)\cos\theta_3} \times \frac{n_1}{n_3} \times \left(\frac{n_3(\omega)}{n'(\omega)}\right)^2 \tag{5.6}$$

where *n*' is the refractive index of the interfacial layer of polymer/N_2 or polymer/water. In Eqs. 5.3–5.6, θ_i is the incidence angle in the medium *i*. When the IR frequency is near vibrational resonance, $\chi^{(2)}_{\mathrm{eff}}$ can be formulated as [4]

$$\chi^{(2)}_{eff} = \chi^{(2)}_{NR} + \sum_q \frac{A_q}{\omega_{IR} - \omega_q + i\Gamma_q} \tag{5.7}$$

where $\chi^{(2)}_{\mathrm{NR}}$ arises from the nonresonant background contribution. Here, A_q, ω_q, and Γ_q are the strength, resonant frequency, and damping coefficient of the *q*th vibrational mode, respectively. Each peak on an SFG spectrum was fitted by Eq. 5.7 using A_q, ω_q, and Γ_q as fitting parameters to find the peak intensity, I^{SFG}.

5.3 Peak Assignments [11]

Fourier-transform infrared spectroscopy (FT-IR) is helpful to assign peaks on an SFG spectrum. If partially deuterated samples are available, they are significantly useful for the peak assignments in the C–H stretching region because SFG peaks originated from C-D bonds appear in a different wavenumber range from the C–H one. In this section, the peak identification of the SFG spectrum for PMMA is discussed with the aid of the corresponding partially deuterated PMMAs.

Figure 5.2 shows the chemical structures of various PMMAs and their SFG spectra over the wavenumber region from 2850 to 3050 cm^{-1} for the various PMMA films in N_2 with the *ssp* and *ppp* polarization combinations. Since this wavenumber region corresponds to the C–H stretching region, no peaks were observed for d_8-PMMA. In the case of d_5-PMMA, C–H bonds are only present in ester methyl groups. While an intense peak was also observed for this compound at 2955 cm^{-1} with the *ssp* combination, an additional weak peak was found at 2990 cm^{-1} with the *ppp* combination. Thus, the SFG peaks at 2955 and 2990 cm^{-1} can be definitely assigned to the C–H symmetric and antisymmetric stretching vibrations of the ester methyl groups.

In the case of d_3-PMMA, C–H bonds are present both in ester methyl and methylene groups. For this compound, an additional peak was observed at 2908 cm^{-1} for the d_3-PMMA film with the *ssp* and *ppp* combinations. Thus, there is no doubt that the peak at 2908 cm^{-1} can be assigned to the C–H symmetric stretching vibrations of the methylene groups. An overlapping peak around 2930 cm^{-1} arising from the antisymmetric stretching of methylene groups might be present in both the *ssp* and *ppp* spectra. The *ppp* peak intensity at 2990 cm^{-1} for d_5-PMMA was similar to that

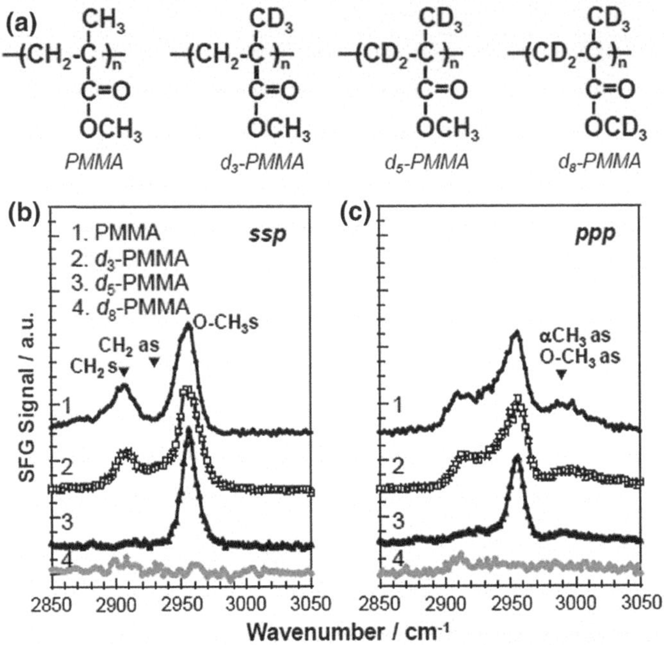

Fig. 5.2 **a** Chemical structures of various PMMAs and SFG spectra for their PMMA films under N_2 atmosphere. **b** *ssp* and **c** *ppp* polarization combinations (Reproduced from Ref. [11] with permission from The Royal Society of Chemistry)

for d_3-PMMA. This is quite reasonable because the peak arises from the C–H antisymmetric stretching vibrations of the ester methyl groups, which are present in both d_5-PMMA and d_3-PMMA. Only PMMA possesses protonated α-methyl groups. The *ssp* spectra of PMMA and d_3-PMMA seem to be identical. On the other hand, the *ppp* spectrum of PMMA is slightly different from that of d_3-PMMA in that the spectral intensity at 2990 cm^{-1} for PMMA was more intense. Taking into account that the structural difference between PMMA and d_3-PMMA is only the presence of α-methyl groups in the former, the above result implies that the peak observed at 2990 cm^{-1} in the *ppp* spectrum involves the C–H antisymmetric stretching of the α-methyl groups. These assignments were in good agreement with those found from FT-IR using bulk films of various PMMAs.

5.4 Orientation of Functional Groups

SFG possesses an advantage for detailed analysis of orientational order of probed molecules. The situation originates from the fact that the C–H vibrational modes are well isolated from other vibrational modes of the molecule [12], and their normal coordinates consist predominantly of the internal symmetry coordinates of the C–H stretching motion with no significant mixing of other internal coordinates. The formulas for the vibrational resonance with the C–H stretching vibrations are thus transferable among molecules.

The macroscopic sum frequency susceptibility tensor $\chi^{(2)}_{\mathrm{ijk}}$ is related to the microscopic hyperpolarizability tensor elements $\beta^{(2)}_{\mathrm{i'j'k'}}$ in the molecular coordinate system defined by *a, b,* and *c* axes. The nonvanishing independent elements in the symmetric stretch hyperpolarizability tensor are extracted based on the symmetry of C–H functional groups including CH_3, CH_2, CH and phenyl groups. The components of $\chi^{(2)}$ are calculated using the nonvanishing hyperpolarizability tensor elements. Furthermore, functional groups existed at the surface possess an orientation angle with a certain fluctuation in general. In this section, it is shown how the orientation of functional groups at interfaces can be estimated.

5.4.1 PMMA/N$_2$ Interface

First, the SFG spectra for the d_5-PMMA film, curves 3 in the panels (b) and (c) of Fig. 5.2, are used to extract the orientation of the ester methyl groups at the N_2 interface. For the *ssp* combination, only the symmetric stretching vibration was observed, qualitatively indicating that the ester methyl groups orient almost totally along the direction perpendicular to the surface. On the other hand, the peak for the antisymmetric stretching vibration of the ester methyl groups was seen at 2990 cm^{-1} only for the *ppp* combination. The tilt angle (θ) of the ester methyl groups can be estimated following a method proposed by Hirose et al. [13–17] namely, using the intensity ratio of the SFG peaks of the symmetric (I_s) to antisymmetric stretching modes (I_{as}). The tilt angle is the one between the *c* and *z* axes. The *z* axis is already defined and is along the interface normal in the laboratory coordinate system, and is hence always fixed. On the other hand, since the *c* axis is in the molecular coordinate system, its direction depends on the molecular symmetry. Thus, the definition of this axis will vary for different functional groups.

Both α-methyl and ester methyl groups can be treated as having C_{3v} symmetry [13, 14]. In this case, there are two nonvanishing independent elements in the symmetric stretch hyperpolarizability tensor: β_{ccc} and $\beta_{aac} = \beta_{bbc}$ [14]. In the molecular-fixed coordinates for a methyl group, the *c* axis coincides with the C_3 axis, and the *a* axis lies in the plane of symmetry σ_v of the methyl group. For the C–H symmetric stretching vibrations, the components of $\chi^{(2)}$ for an isotropic surface are given by [3, 13]

$$\chi_{xxz,s} = \chi_{yyz,s} = N_s\beta_{ccc}\left[\langle\cos\theta\rangle(1+r) - \langle\cos^3\theta\rangle(1-r)\right]/2 \tag{5.8}$$

$$\chi_{xzx,s} = \chi_{yzy,s} = \chi_{zxx,s} = \chi_{zyy,s} = N_s\beta_{ccc}\left[\langle\cos\theta\rangle - \langle\cos^3\theta\rangle(1-r)\right]/2 \tag{5.9}$$

$$\chi_{zzz,s} = N_s\beta_{ccc}\left[r\langle\cos\theta\rangle - \langle\cos^3\theta\rangle(1-r)\right] \tag{5.10}$$

where N_s is the surface density of molecules and $r = \beta_{aac}/\beta_{ccc}$. The r value for the methyl groups of PMMA was given as 1.8 [18]. The brackets < > denote the average over polar angles. For the antisymmetric stretch, the components of $\chi^{(2)}$ for an isotropic surface are expressed as follows [13, 17].

$$\chi_{xxz,as} = \chi_{yyz,as} = -N_s\beta_{caa}\left(\langle\cos\theta\rangle - \langle\cos^3\theta\rangle\right) \tag{5.11}$$

$$\chi_{xzx,as} = \chi_{yzy,as} = \chi_{zxx,as} = \chi_{zyy,as} = N_s\beta_{caa}\langle\cos^3\theta\rangle \tag{5.12}$$

$$\chi_{zzz,as} = 2N_s\beta_{caa}\left[\langle\cos\theta\rangle - \langle\cos^3\theta\rangle\right] \tag{5.13}$$

Generally, functional groups existed at the surface possess an orientation angle with a certain fluctuation. Here, the angle fluctuation was modeled by a Gaussian distribution [19, 20].

$$\langle\cos^n\theta\rangle = \int_0^\pi \cos^n\theta \times f(\theta)\sin\theta d\theta \tag{5.14}$$

$$f(\theta) = C\exp\left[-(\theta-\theta_0)^2/2\sigma^2\right] \tag{5.15}$$

where θ_0 is the average over tilt angles.

Using these equations the orientation of the ester methyl groups at the N_2 interface can be determined. Figure 5.3 shows simulated values of I_{as}/I_s as functions of the average tilt angle θ_0 and the angle distribution σ for the *ssp* and *ppp* polarization combinations. The simulation used Eqs. 5.1–5.6 and Eqs. 5.8–5.15 assuming $\beta_{ccc} \approx \beta_{caa}$ [21] and a value of n' of 1.18 [18]. To find I_{as}/I_s from the experiment, the SFG spectra for d_5-PMMA were separated into two components by curve-fitting using Eq. 5.7. The ratios for the *ssp* and *ppp* combinations were 0 and 0.22, respectively. Comparing these values with the simulation results shown in Fig. 5.3, it seems most likely that θ_0 for the ester methyl groups is approximately 20° with zero angular distribution or that θ_0 is smaller than 15° with an angular distribution less than 15°. In either case, this result indicates that the ester methyl groups at the N_2 interface are well ordered and orient along the direction normal to the interface.

The methylene groups are treated as having C_{2v} symmetry [14]. In the case of C_{2v}, there are three nonvanishing independent elements in the symmetric stretch hyperpolarizability tensor: β_{aac}, β_{bbc}, and β_{ccc}. In the molecular-fixed coordinates for methylene, the *c* axis is defined as the direction along the bisector line of the two C–H bonds, the *a* axis is in the H–C–H plane, and *b* is out of plane. The components of $\chi^{(2)}$ are obtained in the same way as the ester methyl and are described in detail in the literature [13]. The SFG spectra for the PMMA film were again separated into two components, the symmetric and antisymmetric modes for the methylene groups, by curve-fitting using Eq. 5.7. The intensity ratio of the antisymmetric mode to the symmetric one was 0.06 for the *ssp* polarization combination and 4.48 for

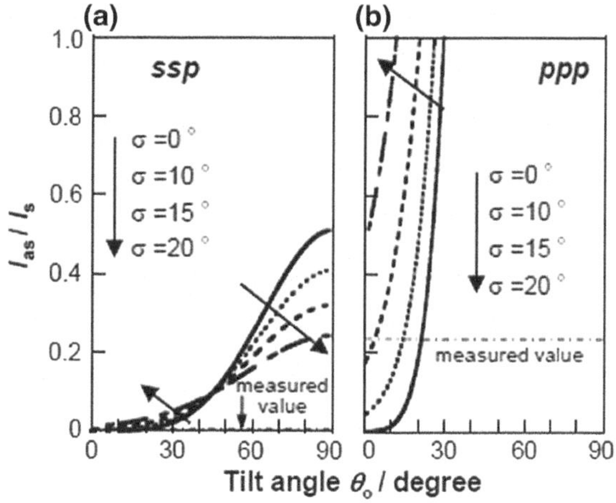

Fig. 5.3 Ratio of the SFG intensity of the C–H antisymmetric and symmetric stretching modes for ester methyl groups of PMMA under N_2 atmosphere as functions of the average tilt angle θ_0 and angular distribution σ. Left and right panels show the plots for **a** *ssp* and **b** *ppp* polarization combinations (Reproduced from Ref. [11] with permission from The Royal Society of Chemistry)

the *ppp* combination. Comparing the experimental I_{as}/I_s values with those from the simulation, it seems most likely that the methylene groups are tilted at about $\theta_0 =$ 20-40° with zero angular distribution or, assuming on a Gaussian distribution, the angular distribution is less than 15°. These results indicate that the main chain of PMMA orients along the direction parallel to the N_2 interface.

Comparing the d_3-PMMA spectra to those of PMMA, the C–H antisymmetric stretching vibration of the α-methyl groups was observed only for the *ppp* combination. Thus, it seems reasonable to infer that the α-methyl groups are oriented almost parallel to the interface.

To summarize this subsection, the hydrophobic methyl and methylene groups were preferentially segregated at the surface, as expected. To achieve this, the main chain part of the PMMA orients in the interfacial plane.

5.4.2 PMMA/Water Interface

Figure 5.4 shows SFG spectra for PMMA films at N_2 and water interfaces. The spectra under N_2 are imported from Fig. 5.2. Interestingly, methylene C–H symmetric stretch at 2908 cm^{-1} disappeared at the water interface. This was a common feature of both *ssp* and *ppp* combinations. Thus, it seems most likely that the methylene groups that were oriented at the N_2 interface became random at the water interface. This can be easily understood if we assume that the side chains of PMMA are segregated to the water phase because of an attractive interaction between carbonyl groups and water molecules.

To confirm whether the above assumption related to the orientation of the carbonyl groups is realistic, SFG spectra for PMMA in the C = O region were acquired before and after the contact with water with the *ssp* polarization combination. Figure 5.5

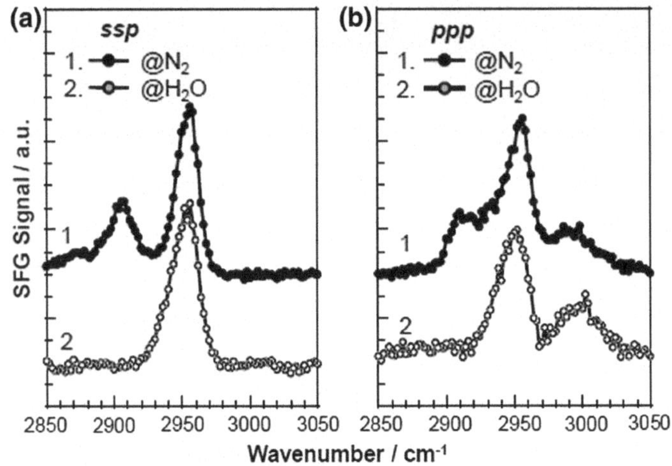

Fig. 5.4 SFG spectra for PMMA at N_2 and water interfaces for **a** *ssp* and **b** *ppp* polarization combinations (Reproduced from Ref. [11] with permission from The Royal Society of Chemistry)

shows the result. At the N_2 interface, a clear peak was observed at 1736 cm^{-1} and could be assigned to free carbonyl groups [22]. The spectrum dramatically changed in shape after contact with water. The peak became broader and shifted to 1714 cm^{-1}. Interfacial hydrogen bonding between the carbonyl groups and water molecules has been discussed in several publications, and the extent of the wavenumber shift was reported to be 10–20 cm^{-1} [6, 23, 24]. The peak may be composed of two overlapping peaks from free and hydrogen bonded carbonyl groups. However, it was impossible to resolve it unambiguously into two peaks. Since the extent of the red shift observed here was 22 cm^{-1}, it seems reasonable to claim that the most of carbonyl groups present at the interface formed hydrogen bonds with water molecules [25].

A SFG measurement for the d_5-PMMA film was performed in deuterated water to gain further insight into the orientation of the PMMA side chains. This experiment probes the orientation of the ester methyl group directly. Figure 5.6 shows that the C–H symmetric stretching vibration of the ester methyl groups was observed in both polarization combinations, whereas the antisymmetric vibration was not observed in either. This means that the ester methyl groups orient at the water interface along the direction perpendicular to the interface. However, the discussion of Fig. 5.5 showed that the carbonyl groups were segregated at the water interface because of an attractive interaction with water molecules. To account for both these observations, the ester methyl groups must orient towards the internal bulk region along the direction normal to the interface. Taking into account the hydrophobicity of the ester methyl groups, such a conformation seems to be reasonable.

For α-methyl groups, Fig. 5.6 also shows that in the case of the d_5-PMMA film in deuterated water, no SFG peaks were observed around 2990 cm^{-1} for both the *ssp* and *ppp* polarization combinations. Similarly, a clear peak was not observed either for the PMMA film in water at the corresponding wavenumber with the *ssp* combination,

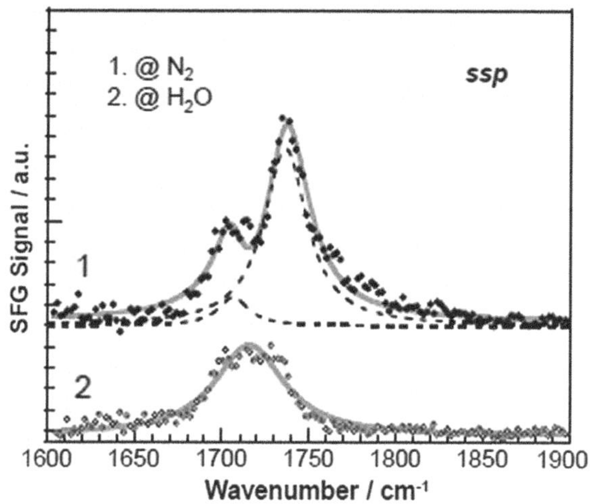

Fig. 5.5 SFG spectra for PMMA at N_2 and water interfaces in the C = O region for the *ssp* polarization combination (Reproduced from Ref. [11] with permission from The Royal Society of Chemistry)

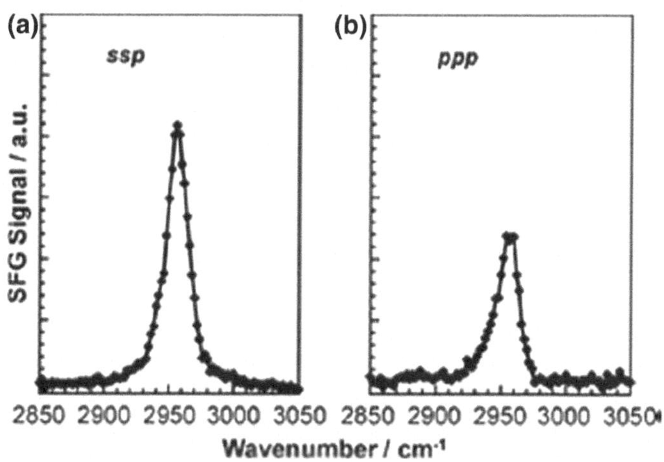

Fig. 5.6 SFG spectra for d_5-PMMA at a deuterated water interface for **a** *ssp* and **b** *ppp* polarization combinations (Reproduced from Ref. [11] with permission from The Royal Society of Chemistry)

as shown by curve 2 in Fig. 5.4a. On the other hand, a clear peak appeared for the PMMA film in water near 2990 cm^{-1} for the *ppp* polarization combination, as shown by curve 2 in Fig. 5.4b. Based on these results, it is conceivable that the α-methyl groups orient parallel to the interface even in the presence of water.

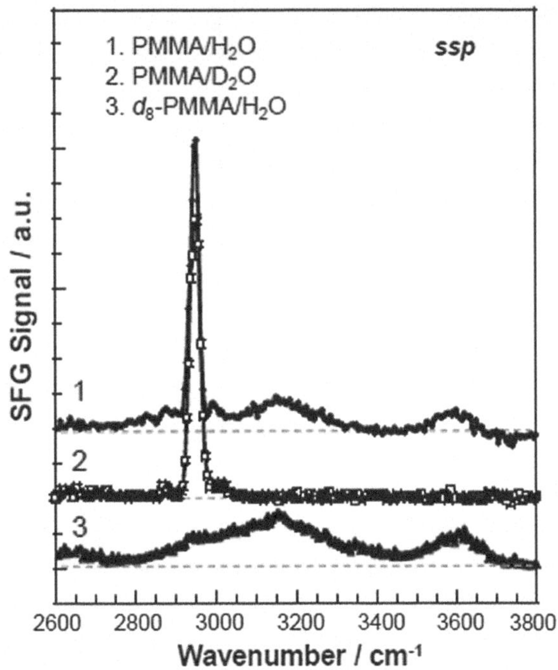

Fig. 5.7 SFG spectra for protonated and deuterated PMMA films at protonated and deuterated water interfaces for the *ssp* polarization combination (Reproduced from Ref. [11] with permission from The Royal Society of Chemistry)

5.4.3 Water Structure at the PMMA Interface

We finally come to the aggregation states of water molecules at the PMMA interface. Figure 5.7 shows SFG spectra for PMMA with water, PMMA with deuterated water and d_8-PMMA with water. In the case of PMMA/water, broad peaks were observed around 3150 and 3600 cm^{-1} in addition to the intense peak corresponding to the C–H symmetric stretching vibration for the ester methyl groups at 2955 cm^{-1}. While the broad peaks at 3150 and 3600 cm^{-1} disappeared for the PMMA film in deuterated water, they again appeared for the d_8-PMMA in water. Therefore, those peaks are assignable to the O–H stretching of water molecules. Comparing our results with the wavenumbers appearing in published reports [5, 26, 27], the peaks at 3150 and 3600 cm^{-1} were specifically assigned to ice-like water and free O–H, respectively.

At the water interface, the outermost region of PMMA might consist of hydrophilic and hydrophobic domains at a sub-nanometer scale. If this is the case, it can be envisaged that water molecules form hydrogen bonding near the hydrophobic domains composed of methylene and α-methyl groups. This leads to the formation of an ice-like structure of water molecules. The peak at 3600 cm^{-1} is a signature of the stretching of only one free O–H of a water molecule, not of two. This is simply because the O–H antisymmetric stretch, which should appear at a wavenumber higher than 3700 cm^{-1} for a free molecule was not observed. Water molecules are supposed to establish hydrogen bonds with their nearest neighbors in a tetrahedral geometry

[26]. However, at the interface, water molecules hydrogen bonded to the carbonyl groups are located in a confined geometry. Such a situation may not allow the water molecules to form a second hydrogen bond, resulting in the presence of free O–H groups.

5.5 Conclusions

In this subsection, SFG vibrational spectroscopy, which is one of the most interfacial sensitive techniques, is introduced. It is first shown that a sample which is in part deuterated is useful to assign SFG peaks. By taking into consideration of the symmetry of the functional groups, the orientation can be determined by the intensity ratio of symmetric to asymmetric stretching modes for both of *ssp* and *ppp* combinations. As an example, the local conformation of PMMA at N_2 and water interfaces was shown. At the N_2 interface, the main chain of PMMA oriented along the direction parallel to the surface. The α-methyl and the ester methyl groups were present at the interface along the directions parallel and normal to the interface, respectively. At the water interface, the PMMA chains in the outermost region of the film reorganized to turn the carbonyl groups towards the water phase to form hydrogen bonds, while the hydrophobic ester methyl groups retreated to the internal bulk phase. However, the α-methyl groups remained in plane at the interface even in the presence of water. The outermost region of PMMA in water probably possesses both hydrophilic and hydrophobic domains at a sub-nanometer scale. On the hydrophobic domains, water molecules form an ice-like network structure without interacting with the PMMA. In contrast, in the hydrophilic regions, they form hydrogen bonds with carbonyl groups. Since the SFG has the potential to give the detail information about the molecular pictures at interfaces, it will contribute to understanding important factors to control "interfacial engineering" such as adhesion, tribology, etc.

References

1. Zhu XD, Suhr H, Shen YR (1987) Surface vibrational spectroscopy by infrared-visible sum frequency generation. Phys Rev B 35(6):3047–3050. https://doi.org/10.1103/PhysRevB.35.3047
2. Shen YR (1989) Surface-properties probed by 2nd-harmonic and sum-frequency generation. Nature 337(6207):519–525. https://doi.org/10.1038/337519a0
3. Zhuang X, Miranda PB, Kim D, Shen YR (1999) Mapping molecular orientation and conformation at interfaces by surface nonlinear optics. Phys Rev B 59(19):12632–12640. https://doi.org/10.1103/PhysRevB.59.12632
4. Chen Z, Shen YR, Somorjai GA (2002) Studies of polymer surfaces by sum frequency generation vibrational spectroscopy. Annu Rev Phys Chem 53:437–465. https://doi.org/10.1146/annurev.physchem.53.091801.115126

5. Richmond GL (2002) Molecular bonding and interactions at aqueous surfaces as probed by vibrational sum frequency spectroscopy. Chem Rev 102(8):2693–2724. https://doi.org/10.1021/cr0006876
6. Chen Z (2007) Understanding surfaces and buried interfaces of polymer materials at the molecular level using sum frequency generation vibrational spectroscopy. Polym Int 56(5):577–587. https://doi.org/10.1002/pi.2201
7. Lu X, Shephard N, Han J, Xue G, Chen Z (2008) Probing molecular structures of polymer/metal interfaces by sum frequency generation vibrational spectroscopy. Macromolecules 41(22):8770–8777. https://doi.org/10.1021/ma801680f
8. Wilks PA (1962) Infrared analysis by attenuated total reflectance. Spectrochim Acta 18(6):884
9. Chen YJ, Chen WP, Burstein E (1976) Surface-electromagnetic-wave-enhanced Raman-scattering by overlayers on metals. Phys Rev Lett 36(20):1207–1210. https://doi.org/10.1103/PhysRevLett.36.1207
10. Noguchi H, Hiroshi M, Tominaga T, Gong JP, Osada Y, Uosaki K (2008) Interfacial water structure at polymer gel/quartz interfaces investigated by sum frequency generation spectroscopy. PCCP 10(32):4987–4993. https://doi.org/10.1039/b807297n
11. Tateishi Y, Kai N, Noguchi H, Uosaki K, Nagamura T, Tanaka K (2010) Local conformation of poly(methyl methacrylate) at nitrogen and water interfaces. Polym Chem 1(3):303–311. https://doi.org/10.1039/b9py00227h
12. Bellamy LJ (1975) The infra-red spectra of complex molecules, 3d edn. Chapman and Hall, Wiley, London, New York
13. Hirose C, Akamatsu N, Domen K (1992) Formulas for the analysis of the surface SFG spectrum and transformation coefficients of cartesian SFG tensor components. Appl Spectrosc 46(6):1051–1072. https://doi.org/10.1366/0003702924124385
14. Hirose C, Akamatsu N, Domen K (1992) Formulas for the analysis of surface sum-frequency generation spectrum by CH stretching modes of methyl and methylene groups. J Chem Phys 96(2):997–1004. https://doi.org/10.1063/1.462120
15. Akamatsu N, Domen K, Hirose C (1993) SFG study of 2-dimensional orientation of surface methyl-groups on cadmium arachidate Langmuir-Blodgett-films. J Phys Chem 97(39):10070–10075. https://doi.org/10.1021/j100141a029
16. Hirose C, Yamamoto H, Akamatsu N, Domen K (1993) Orientation analysis by simulation of vibrational sum-frequency generation spectrum - CH stretching bands of the methyl-group. J Phys Chem 97(39):10064–10069. https://doi.org/10.1021/j100141a028
17. Watanabe N, Yamamoto H, Wada A, Domen K, Hirose C, Ohtake T, Mino N (1994) Vibrational sum-frequency generation (VSFG) spectra of N-alkyltrichlorosilanes chemisorbed on quartz plate. Spectrochim Acta Part A Mol Biomol Spectrosc 50(8–9):1529–1537. https://doi.org/10.1016/0584-8539(94)e0064-h
18. Wang J, Chen CY, Buck SM, Chen Z (2001) Molecular chemical structure on poly(methyl methacrylate) (PMMA) surface studied by sum frequency generation (SFG) vibrational spectroscopy. J Phys Chem B 105(48):12118–12125. https://doi.org/10.1021/jp013161d
19. Simpson GJ, Rowlen KL (1999) An SHG magic angle: dependence of second harmonic generation orientation measurements on the width of the orientation distribution. J Am Chem Soc 121(11):2635–2636. https://doi.org/10.1021/ja983683f
20. Wei X, Zhuang XW, Hong SC, Goto T, Shen YR (1999) Sum-frequency vibrational spectroscopic study of a rubbed polymer surface. Phys Rev Lett 82(21):4256–4259. https://doi.org/10.1103/PhysRevLett.82.4256
21. Ye S, Morita S, Li GF, Noda H, Tanaka M, Uosaki K, Osawa M (2003) Structural changes in poly (2-methoxyethyl acrylate) thin films induced by absorption of bisphenol A. An infrared and sum frequency generation (SFG) study. Macromolecules 36 (15):5694–5703. https://doi.org/10.1021/ma026007d
22. Fleischer CA, Morales AR, Koberstein JT (1994) Interfacial modification through end group complexation in polymer blends. Macromolecules 27(2):379–385. https://doi.org/10.1021/ma00080a010

23. Li GF, Ye S, Morita S, Nishida T, Osawa M (2004) Hydrogen bonding on the surface of poly(2-methoxyethyl acrylate). J Am Chem Soc 126(39):12198–12199. https://doi.org/10.1021/ja046183x
24. Chen CY, Clarke ML, Wang J, Chen Z (2005) Comparison of surface structures of poly (ethyl methacrylate) and poly (ethyl acrylate) in different chemical environments. PCCP 7(11):2357–2363. https://doi.org/10.1039/b501910a
25. Wan L-S, Huang X-J, Xu Z-K (2007) Diffusion and structure of water in polymers containing N-vinyl-2-pyrrolidone. J Phys Chem B 111(5):922–928. https://doi.org/10.1021/jp065152g
26. Du Q, Superfine R, Freysz E, Shen YR (1993) Vibrational spectroscopy of water at the vapor water interface. Phys Rev Lett 70(15):2313–2316. https://doi.org/10.1103/PhysRevLett.70.2313
27. Du Q, Freysz E, Shen YR (1994) Surface vibrational spectroscopic studies of hydrogen-bonding and hydrophobicity. Science 264(5160):826–828. https://doi.org/10.1126/science.264.5160.826

Chapter 6
Surface Analysis

Daisuke Kawaguchi and Keiji Tanaka

6.1 X-Ray Photoelectron Spectroscopy (XPS)

6.1.1 Introduction

X-ray photoelectron spectroscopy (XPS), which is one of the most surface-sensitive spectroscopy, provides much information of which kind of atoms is existed at the surface of materials. In addition, chemical states of the atoms can be also analyzed.

The technique is based on the principle that photoelectrons are emitted when a sample is irradiated with soft X-rays such as MgKα or AlKα. This is the so-called photoelectric effect. The photoelectrons emitted are collected by a lens system and are focused into an energy analyzer. The number of electrons is counted with given kinetic energy (E_K). The binding energies (E_B) of the photoelectrons are obtained by using the following Einstein equation;

$$E_B = h\nu - E_K - \Phi \tag{6.1}$$

where $h\nu$ is the photon energy of X-rays and Φ is the work function of the sample.

All elements except hydrogen and helium can be detected in principle and the core electron binding energies are characteristic of the atomic core levels. The core-binding energies in an atom are also influenced by the local electronic environment, and consequently, an atom in a molecule can exhibit a small range of binding energies—known as chemical shifts. For example, oxygen induces shifts in the binding energy of carbon by approximately 1.5 eV per C–O bond, while halogens induce shifts to higher binding energies for carbon in the range 1–3 eV. The chemical shifts

D. Kawaguchi · K. Tanaka (✉)
Department of Applied Chemistry, Kyushu University, Fukuoka 819-0395, Japan
e-mail: k-tanaka@cstf.kyushu-u.ac.jp

M. Maeda et al. (eds.), *Molecular Soft-Interface Science*,
https://doi.org/10.1007/978-4-431-56877-3_6

in XPS are similar to those in a spectrum of nuclear magnetic resonance (NMR), and thus, enable us to determine the surface chemical composition of multicomponent soft materials.

6.1.2 Surface Segregation in Polymer Blends [1]

When two different chemical species are mixed, one of them is enriched at the surface. This phenomenon, so-called surface segregation, can take place even in a binary miscible system. Controlling the surface chemical composition in soft materials using the segregation phenomenon is a useful method to functionalize the surface of materials without changing any properties in the bulk. Here, we show the XPS analysis of the surface segregation in a blend of poly(2-methoxyethyl acrylate)(PMEA) and poly(methyl methacrylate)(PMMA). PMEA is a new bio-inert and partially water-soluble polymer and PMMA is a water-insoluble polymer.

PMEA with a number-average molecular weight (M_n) of 26 k and a molecular weight distribution (M_w/M_n) of 3.23, where M_w denotes a weight-average molecular weight, was synthesized by radical polymerization [2]. Monodisperse PMMA with M_n of 85 k and M_w/M_n of 1.09 was purchased from Polymer Source Inc. The surface free energies (γ) in air of PMEA and PMMA were determined to be 36.7 and 42.2 mJ m^{-2}, respectively, by contact angle measurements according to Owens' procedure [3].

The surface chemical composition of the PMEA/PMMA blend films was examined by X-ray photoelectron spectroscopy (XPS, PHI 5800 ESCA system, Physical Electronics, Co., Ltd) with a monochromatized AlKα source operated at 14 kV and 24 mA. The pressure in the main chamber of XPS was maintained at ~10^{-8} Pa or even better. The C_{1s} peak was calibrated to a binding energy of 285.0 eV for neutral carbon, to correct the charging energy shifts.

To extract the depth profile of chemical composition using XPS, there are two ways, angular-dependent measurement, and ion-beam etching technique. The latter is that the sample is etched by ion beams and then the conventional XPS measurement is carried out. Although this possesses an intriguing advantage that the information in deeper region is accessible, the ion beams strongly damage to the sample surface. Thus, this may not be suitable for soft materials. However, it has been recently reported that a fullerene (C_{60}) ion beam causes an extremely low sputtering damage to polymer samples in comparison with monatomic ion beams such as argon, xenon, and neon ion beams [4]. The sputtering technique using C_{60} ion beam will be one of the keys to the development of XPS analyses for soft materials in deeper region.

Alternatively, it is possible to examine the depth profile of a component changing take-off-angles (θ) of photoelectrons to the detector. In this case, the maximum value of the analytical depth (d) is limited to approximately 10 nm. However, since the measurement itself is not invaded the sample at all, this way is very useful for soft materials, as seen in the next.

The d value is defined by

$$d = 3\lambda \sin\theta \tag{6.2}$$

where λ is the inelastic mean-free path of the photoelectrons in the solids. The inelastic mean-free path of C1 s photoelectrons was taken as 3.1 nm, calculated by Ashley's equation [5]. The depth profile of PMEA composition in the vicinity of the blend film surface was analyzed by angular-dependent XPS [6] over a range of θ from 15 to 90°.

Figure 6.1 shows XPS C1 s spectra obtained at $\theta = 45°$ for the PMEA/PMMA blend films with various bulk blend ratios before the annealing treatment. The C1 s peaks corresponding to neutral, ether, and carbonyl carbons were observed at 285.0, 286.5, and 289.0 eV, respectively. Photoelectron intensity of ether carbons increased with increasing PMEA blend ratio because PMEA contains more ether carbons than PMMA. The surface PMEA weight fraction (w^{s}_{PMEA}) can be calculated by the following equation:

$$\frac{I_{C-O}}{I_{C_{total}}} = \frac{3w^{s}_{PMEA}/M_{MEA} + w^{s}_{PMMA}/M_{MMA}}{6w^{s}_{PMEA}/M_{MEA} + 5w^{s}_{PMMA}/M_{MMA}} \tag{6.3}$$

where I_i is the integrated intensity of the peak for a component i. M_{MEA} and M_{MMA} are the molecular weights of 2-methoxyethyl acrylate and methyl methacrylate units, respectively.

Figure 6.2 shows the surface chemical composition of the PMEA/PMMA blend films before and after annealing at 413 K. The surface PMEA fraction is higher than the bulk over the entire range of blend ratios. Considering the lower surface energy as well as the lower degree of polymerization of PMEA compared to those

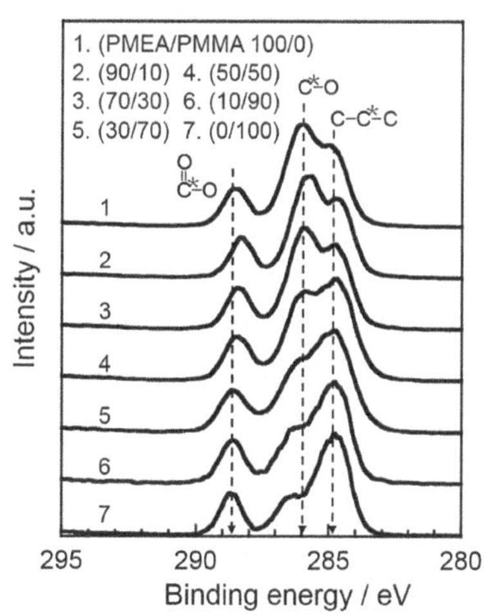

Fig. 6.1 XPS C1 s core level spectra of PMEA/PMMA blend thin films with various bulk ratios at $\theta = 45°$. All neutral C1 s peaks were assigned a binding energy of 285.0 eV to correct for the charging energy shift (Reproduced from Ref. [1] with permission from The Royal Society of Chemistry)

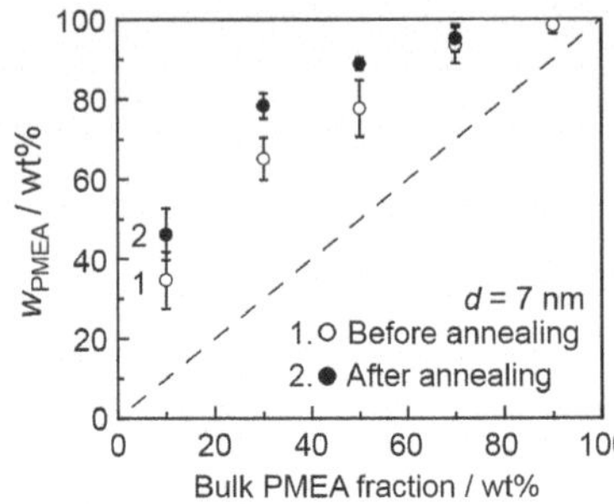

Fig. 6.2 Relation between bulk and surface fractions for PMEA/PMMA films before and after annealing at 413 K. The broken line denotes that the surface and bulk compositions are the same (Reproduced from Ref. [1] with permission from The Royal Society of Chemistry)

of PMMA, namely, the enthalpic and entropic advantages of PMEA, the surface segregation of PMEA is quite reasonable. The surface fraction of PMEA increased after the annealing treatment. This means that the surface of the as-prepared films was not in a quasi-equilibrium state.

Since the θ for Fig. 6.2 was 45°, the analytical depth was roughly calculated to be 7 nm by Eq. 6.2. Thus, the surface PMEA fraction is averaged over this depth range. To extract the depth profile of PMEA in the vicinity of the surface of the blend, an ADXPS measurement was made, as introduced above. Figure 6.3a shows the surface PMEA volume fraction (φ^{s}_{PMEA}) in the PMEA/PMMA (10/90 w/w) blend films as a function of sin θ, which is proportional to the analytical depth. The film was annealed under the same condition as before so that the surface can reach a quasi-equilibrium state. Cloud point measurements revealed that the PMEA/PMMA (10/90) blend was in a miscible state. The volume fraction is used here instead of the weight fraction because the fitting to the sin θ versus φ^{s}_{PMEA} relation on the basis of a mean-field prediction becomes easier. For the blend film, φ^{s}_{PMEA} increased with decreasing sin θ, meaning that the PMEA component was enriched toward the surface.

In general, photoelectrons cannot travel for a long distance in a solid due to inelastic scattering. This means that only photoelectrons emitted from the region in close proximity to the surface can get out of the solid, resulting in the surface sensitivity of the XPS technique. The photoelectron intensity for the j-core level at θ is expressed as [7, 8]

$$I_j(\theta) = Fk \int_0^{\infty} n_j(z) \exp\left\{\frac{-z}{\lambda_j \sin\theta}\right\} dz \tag{6.4}$$

where z and $n_j(z)$ represent depth and atomic composition-depth profile, respectively. F and k are the transmission function and a factor related to sensitivity, respectively. Hence, even though the analytical depth is z, photoelectrons are not uniformly emitted from the depth region from the surface to z. Instead, the detected amount of photoelectrons exponentially decays with increasing depth to z. This means that the dependence of surface composition on sin θ, Fig. 6.3a, cannot be simply regarded as the compositional depth profile. Thus, the following treatment was made to extract a plausible composition profile near the surface in the real space.

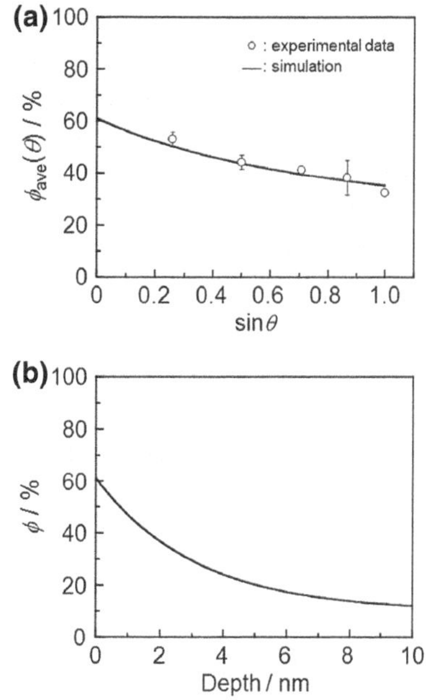

Fig. 6.3 **a** Sin θ dependence of PMEA volume fraction for the PMEA/PMMA (10/90) blend films annealed at 413 K for 3 h. Circles are the experimental data, and solid line denotes the best fit to Eq. 6.6 in the text. **b** Model profiles of $\phi^{s}_{PMEA}(z)$ to fit the sin θ versus $\phi^{s}_{PMEA}(\theta)$ relation in panel (**a**) (Reproduced from Ref. [1] with permission from The Royal Society of Chemistry)

Schmidt and Binder, using a mean-field approximation, proposed that the surface composition profile ($\varphi^{s}(z)$) in a miscible polymer blend can be given by the following equation [9]:

$$\varphi^{s}(z) = \varphi_{\infty} + \left(\varphi^{s}_{o} - \varphi_{\infty}\right) \exp\left(\frac{-z}{\xi}\right) \tag{6.5}$$

where φ^{s}_{0} and φ_{∞} are the surface ($z = 0$) and bulk volume fractions of a component, respectively, and ξ is the decay length showing how the surface composition reaches the bulk value. Thus, the composition of the PMEA/PMMA (10/90) blend film by XPS at a given θ can be expressed by

$$\varphi^{s}(\theta) = \frac{\int_0^{\infty} \varphi^{s}(z) \exp\left\{\frac{-z}{\lambda_j \sin\theta}\right\} dz}{\int_0^{\infty} \exp\left\{\frac{-z}{\lambda_j \sin\theta}\right\} dz} \tag{6.6}$$

Accordingly, the sin $\theta - \varphi^{s}_{PMEA}(\theta)$ curve was fitted with the φ^{s}_{0} and ξ parameters in Eq. 6.5, given that φ_{∞} was set to be the initial mixing ratio. When the $\varphi^{s}_{PMEA}(z)$ shown in panel (b) of Fig. 6.3 is assumed, the best fit to the experimental data is shown by the solid curve in Fig. 6.3a. In the case of the PMEA/PMMA (10/90) blend film, φ^{s}_{0} and ξ were estimated to be 0.61 and 3.5 nm, respectively.

6.1.3 Chain End Segregation in Polymer Blends [6]

As shown in the previous part, the lower γ component is segregated at the surface in a binary polymer blend. However, if a polymer possesses chain end groups with a significantly low γ, they are segregated at the surface. Here, we present the chain end segregation at the surface in a blend of polystyrene with fluoroalkyl end groups at both ends (α,ω-PS(R_f)$_2$) and poly(vinyl methyl ether)(PVME). If PS without chain end modification is used, PVME is segregated at the surface of PS/PVME blend film because the γ of PVME is lower than that of PS. However, the γ of R_f end group used in this study is much lower than those of both PVME and the main chain part of PS.

Figure 6.4 shows the analytical depth dependence of peak intensity ratio of F1 s to C1 s, I_{F1s}/I_{C1s}, in the α,ω-PS(R_f)$_2$/PVME blend films as a function of degree of polymerization (N) to verify an effect of chain end concentration on the surface segregation. Since F atom is contained only in the chain end portion, the I_{F1s}/I_{C1s} value can be regarded as an indicator of the chain end concentration. For all α,ω-PS(R_f)$_2$/PVME blend films employed, the I_{F1s}/I_{C1s} value increased with decreasing $\sin\theta$. This result clearly indicates that the R_f end groups are preferentially partitioned to the surface. However, the I_{F1s}/I_{C1s} versus $\sin\theta$ relation cannot be directly referred to as the true depth profile, as mentioned above. Hence, the true depth profile was obtained on the basis of Paynter's algorithm [7]. The solid curves in Fig. 6.4 depict the best-fit relations of I_{F1s}/I_{C1s} to $\sin\theta$ using models shown in the inset. On the basis of the inset of Fig. 6.4, it is envisaged that even in the α,ω-PS(R_f)$_2$/PVME blend films, the R_f end groups are almost perfectly localized at the surface due to the lowest surface energy in the system.

Since the PS segments are directly connected to the R_f groups, they are inevitably pulled out to the surface. This is as though the chain ends behave like floating buoys

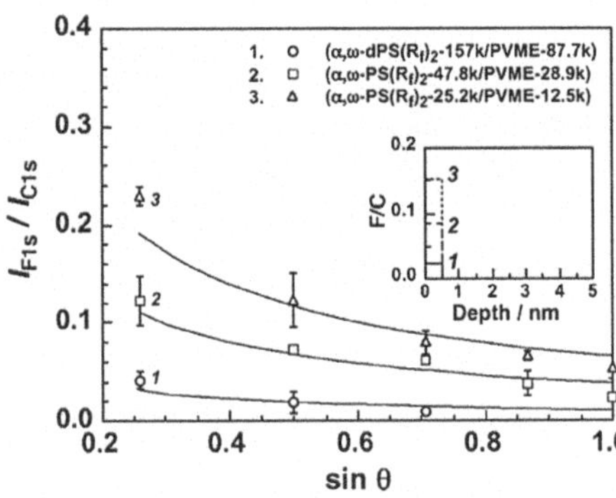

Fig. 6.4 Relation between $\sin\theta$ and intensity ratio of F1 s to C1 s for α,ω-PS(R_f)$_2$/PVME blend films with different M_n pairs. Open symbols are the experimental data sets, and solid curves are the best-fit ones calculated on the basis of model depth profiles of (F/C) shown in the inset (*Macromolecules*, **36**, 6824 (2003). Copyright © 2003, American Chemical Society)

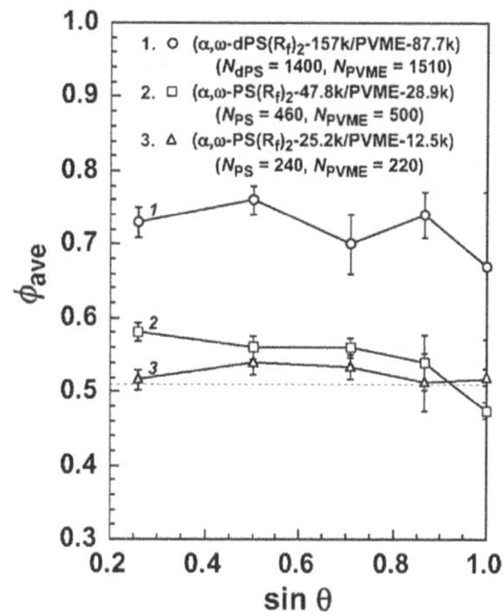

Fig. 6.5 Sin θ dependence of surface PVME fraction for symmetric α,ω-PS(R_f)$_2$/PVME blend films with different M_n pairs (*Macromolecules*, **36**, 6824 (2003). Copyright © 2003, American Chemical Society)

for the PS segments. If this notion is correct, to what extent the PVME component is enriched at the surface should be inversely proportional to the number density of the R_f chains ends, namely N. This is actually what was observed in Fig. 6.5. Hence, it seems reasonable to conclude that the PVME fraction at the surface in the α,ω-PS(R_f)$_2$/PVME films was suppressed, owing to the surface localization of the R_f chain ends. The results presented here make it clear that surface chemical composition in polymer blends can be somehow controlled by chain end chemistry of a component.

6.1.4 Summary

In this subsection, the characteristics of XPS analyses on polymer blends are briefly reviewed. We show that the depth profile of a component in soft materials can be determined from the angular-dependent measurement.

6.2 Secondary Ion Mass Spectroscopy (SIMS)

6.2.1 Introduction

The principle of SIMS is literally. That is, it is based on the mass spectroscopic analysis of atoms or molecular fragments positively and negatively charged, which are generated by the interaction of a primary ion beam (keV range) with a solid sample [10]. SIMS can be roughly categorized into two modes; static and dynamic.

In the case of static SIMS (SSIMS), the amount of ion dose during the measurement is low enough for the surface to be strongly damaged. This is like that the outermost layer at the surface is peeled off one by one. This leads to that SSIMS can provide useful information on identifying the composition of soft materials. In most cases, a time-of-flight detector is used to measure mass for the static mode. Thus, SSIMS is commonly designated as TOF SIMS.

In contrast, abundant primary ions are irradiated onto the sample surface for dynamic SIMS (DSIMS). Consequently, atomic ions are generated from the sample. The high current density of the primary beams allows the fast erosion of a sample so that the depth profile of components can be extracted. In this subsection, we show the surface chemical composition of isotopic polymer blends examined by SSIMS and the time evolution of interfacial thickness with annealing examined by DSIMS.

6.2.2 Surface Composition in an Isotopic Polymer Blend (SSIMS) [11]

The small difference in polarizability between C–H and C–D bonds leads to a very small difference in surface energy. Even in this case, the surface segregation can be attained. Besides, a lower molecular weight component is preferentially segregated at the film surface of an asymmetric polymer blend, in which the components possess different N. This is because shorter chains would suffer less of an entropic penalty by being near a surface. Here we first show the surface composition in films of an isotopic polymer blend composed of normal polystyrene (hPS) and deuterated polystyrene (dPS) examined by SSIMS.

As samples, hPS with M_n of 19.7 k (hPS-19.7 k) and dPS with M_n of 847 k (dPS-847 k) were used. For SSIMS measurements, Phi TRIFT II with a time-of-flight mass spectrometer was used. The sample surface with an area of $100 \times 100\ \mu m^2$ was irradiated by a Ga pulse beam of 15 keV with a current of 600 pA. The pulse width was 13 ns and was generated 10^4 times/s. The spectra integration was allowed for 3 min.

Figure 6.6 shows a typical SSIMS spectrum of the (hPS-19.7 k/dPS-847 k) film with the bulk composition of (41.5/58.5) in volume. The intense peaks observed at the mass of 91.05 and 98.10 Da can be assigned to tropylium ion, $C_7H_7^+$, and its deuterated species, $C_7D_7^+$, respectively. Also, the peaks at 92.06 and 97.09 Da are due probably to $C_7DH_6^+$ or $C_6C^{13}H_7^+$ and $C_7D_6H^+$. While the secondary ion intensities at 91.05 and 98.10 Da monotonically increased and decreased with increasing feed fraction of PS into the blend, respectively, those at 92.06 and 97.09 Da were not necessarily proportional to the blend ratio. Hence, the surface hPS fraction in the blend was estimated from the value of $I_{91}/(I_{91} + I_{98})$, where I_i is the secondary ion intensity at i Da, provided that a contribution of chain end fragments to I_{91} was corrected by following the procedure of Eynde et al. [12]. Also, the $(I_{91} + I_{98})/(I_{91} + I_{92} + I_{97} + I_{98})$ value was in accordance with the value of $I_{91}/(I_{91} + I_{98})$ within

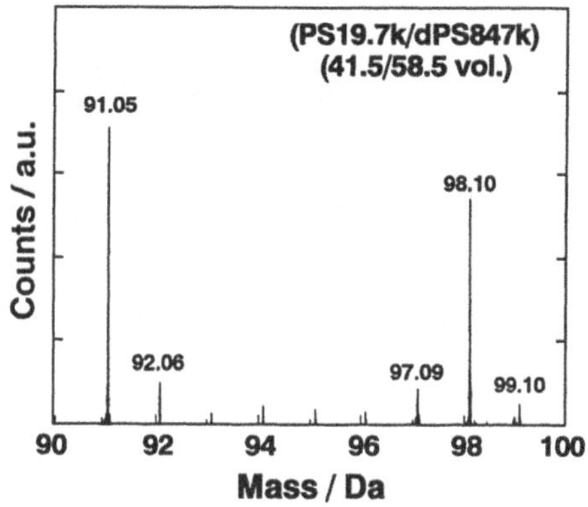

Fig. 6.6 Positive secondary ion time-of-flight mass spectrum of (hPS-19.7 k/dPS-847 k) blend film with the bulk PS fraction of 41.5 vol.% (*Macromolecules*, **35**, 4702 (2002). Copyright © 2002, American Chemical Society)

5%. The surface PS fraction in the (hPS-19.7 k/dPS-847 k) blend films so obtained was 55.5:44.5. The surface segregation of hPS-19.7 k in the (hPS-19.7 k/dPS-847 k) blend films can be explained in terms of an entropic effect.

6.2.3 Mobility Gradient Near Polymer Surfaces (DSIMS) [13]

Many studies have revealed that the surface mobility and dynamics in polymer films are enhanced in comparison with those in the bulk. Based on these, it is reasonable to infer that polymer chains existed in the surface region can diffuse even at a temperature lower than the glass transition temperature in the bulk (T_g^b) as long as the temperature is higher than the T_g at the surface (T_g^s). Perpendicular diffusion in the surface region can be examined by using a bilayer film composed of two different components in which the two original surfaces stand face-to-face. Following an annealing treatment, the bilayer interface is broadened on account of the chain interdiffusion. In this subsection, we show that time evolution of interfacial thickness of hPS/dPS bilayer films studied by DSIMS.

As samples, hPS and dPS with both M_n of 29 k were used. Laminated (hPS/dPS) bilayer films were prepared by a floating method described in detail elsewhere [14]. At first, the bottom dPS layer for the bilayer was coated from a toluene solution onto a silicon wafer by the spin-coating method. The thickness of this layer was approximately 300 nm. The top hPS film with the almost same thickness was independently coated onto a microscope slide glass by a similar manner. Both films were annealed at 393 K for at least 36 h in vacuo to remove the residual solvent and the strain imposed by the film preparation process. The perimeter of the hPS film was scored with a blade, and the film was successively floated off onto the surface of 2.7 wt% 1-hydro-2-fluoroammonium solution. Then, the hPS film was picked up onto the dPS film by attaching the dPS film from the air side, resulting in that the bilayer interface was built up by two original surfaces of the hPS and dPS films.

For interdiffusion experiment, the bilayers were annealed under nitrogen atmosphere at various temperatures. The oven temperature was set to be the middle between the T_g^s and the T_g^b or well above the T_g^b. Once the inside temperature of the oven reached a constant, the bilayers were stored in the oven. Then, the annealing time was started to count. After a given time, the bilayers were rapidly quenched by immersing them into liquid nitrogen. The temperature calibration was made using mercury thermometers, which can measure the first decimal point. The temperature accuracy was 0.5 K.

Annealing-induced interfacial evolution of the bilayers was examined by DSIMS (SIMS 4000, Seiko Instruments Inc., Atomika Analysetechnik GmbH). To gain access to a stable sputtering during the measurement, the buffer dPS layer was laminated onto the (hPS/dPS) bilayer by the floating technique. The thickness of the buffer dPS layer was approximately 200 nm. The incident beam of oxygen ions with 4 keV and ca. 30 nA was focused onto a 200 μm × 200 μm area of the specimen surface. The incident angle was 45°. A gold layer of 20 nm thickness was sputter-coated on the bilayer surface to avoid charging of the specimen during the DSIMS measurement.

Figure 6.7 shows a typical DSIMS profile of proton, H^+, deuterium, D^+, and carbon ions, C^+, for the (hPS/dPS) bilayer. For the first few minutes, the outermost gold layer was etched, resulting in that secondary ions from the polymers were not clearly detected. After the gold layer, the C^+ intensity started to increase and then remained almost constant through the bilayer. Hence, it can be judged that the steady-state etching proceeded during the measurement. While the D^+ intensity was relatively stronger than the H^+ one in the buffer dPS layer, the intensity relation became opposite in the hPS layer. Then, the relation of D^+ to H^+ intensity was again recovered when the etching reached the bottom dPS layer. Postulating that a constant etching was attained through the bilayer, the abscissa of etching time can be simply converted to the depth from the surface. The etching rate was preexamined using the dPS film with a known thickness. We are interested in the thickness of the (hPS/dPS) bilayer interface marked in Fig. 6.7.

A measured concentration profile by DSIMS is generally broadened from an ideal one owing to an atomic mixing effect. The broadening of the obtained profile was quantified by the instrument function, Δz_g, corresponding to the depth resolution. Figure 6.8 shows our definition of the interfacial thickness: (a) the cartoon showing the bilayer configuration, (b) the deuterium ion intensity profile $I_{D+}(z)$ through the interface, and (c) the derivative of $I_{D+}(z)$ by the distance from the center of the interface. Assuming that the $dI_{D+}(z)/dz$ can be expressed by Gaussian function as shown in Fig. 6.8c, the Δz_i ($i = g$ or m) is defined as twice the standard deviation of Gaussian function, corresponding to the depth range where I_{D+} rises from 16% to 84% of the maximum value [15]. The Δz_m denotes the measured apparent width of the bilayer interface. Using (hPS-190 k/dPS-185 k) bilayer film without annealing treatment, the Δz_g was estimated to be 7.4 nm. Hereafter, the number following each polymer species denotes its M_n. Since the interdiffusion at this bilayer interface cannot take place due to the T_g^s of 335 K being much higher than room temperature

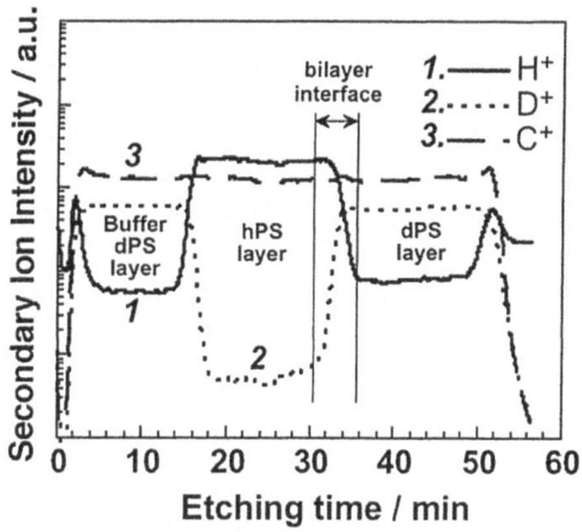

Fig. 6.7 Typical DSIMS profile of proton, H^+, deuterium, D^+, and carbon, C^+, ions for the (hPS-29 k/dPS-29 k) bilayer. For stable experiments, gold and buffer dPS layers were mounted on the bilayer. The interfacial broadening of the bilayer was discussed on the basis of D^+ intensity change (*Macromolecules*, **36**, 1235 (2003). Copyright © 2003, American Chemical Society)

Fig. 6.8 Our definition of the interfacial thickness: **a** cartoon showing bilayer configuration composed of hPS and dPS; **b** deuterium ion intensity profile, $I_{D+}(z)$, through the interface; **c** derivative of $I_{D+}(z)$ by the distance from the center of the interface (*Macromolecules*, **36**, 1235 (2003). Copyright © 2003, American Chemical Society)

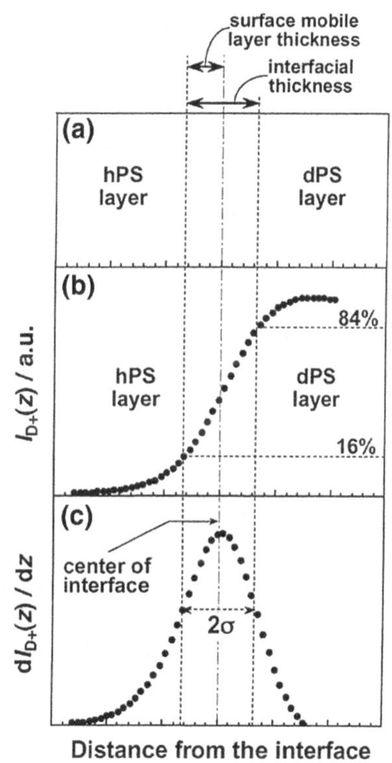

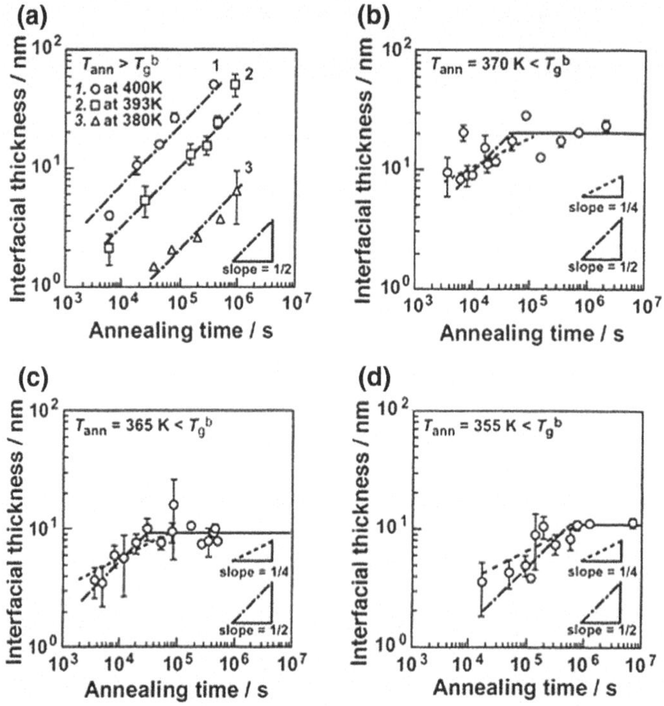

Fig. 6.9 Double-logarithmic plots of the relation between interfacial thickness and annealing time for (hPS-29 k/dPS-29 k) bilayer annealed at various temperatures: **a** 400, 393, and 380 K above both T_g^b and T_g^s; **b** 370 K below T_g^b and above T_g^s; **c** 365 K below T_g^b and above T_g^s; **d** 355 K below T_g^b and above T_g^s. The broken and dotted lines are drawn in the context of Fickian and segmental diffusions, respectively (*Macromolecules*, **36**, 1235 (2003). Copyright © 2003, American Chemical Society)

[16], the measured interfacial width can be regarded as the Δz_g. Then, the real interfacial thickness Δz is given in terms of the Δz_m and the Δz_g [15].

Diffusion behavior of polymers strongly depends on temperature. It was examined whether the interfacial thickness increased even at a temperature below the T_g^b. Parts (a)–(d) of Fig. 6.9 show the time evolution of interfacial thickness for the (hPS-29 k/dPS-29 k) bilayer as a function of temperature. In the case of the annealing at 400, 393, and 380 K being above the T_g^b of 376 K, the interfacial thickness proportionally increased to a half power of the annealing time, t. This is in good accordance with the context of Fickian diffusion. On the contrary, a unique interfacial evolution was observed at 370 K being in between the T_g^s and the T_g^b. At first, the bilayer interface monotonically thickened with increasing t, although the exponent of t could be hardly determined because of the data scattering. When t proceeded to 10^5 s, however, the interfacial thickness remained the constant of 20 ± 5.6 nm. Here, it should be reminded that the bilayer interface was prepared by attaching two original surfaces of hPS and dPS together. Thus, the data mean that chains went across the "mobile" interface and then reached the "dead" bulk region in terms of diffusivity. In other words, half of the constant interfacial thickness evolved after a sufficiently long time would correspond to the surface layer, in which the mobility is enhanced in comparison with the internal bulk phase. We here define this layer as the "surface mobile layer", as shown in Fig. 6.9a. Deferring what the assumption of this notion is, the annealing temperature dependence of such a surface mobile layer is discussed. At 370 K, the thickness of the surface mobile layer was 10 ± 2.8 nm. It should be of interest to compare the thickness with the chain dimension. Twice the radius of gyration, $2R_g$, of an unperturbed PS-29 k is calculated to be 9.3 nm by $2(Nb^2/6)^{1/2}$, where b is the Kuhn's statistical segment length, respectively. This value is comparable to the surface layer thickness. Hence, it seems most likely that the quasi-equilibrium interfacial thickness was attained by center-of-mass diffusion such as Fickian. At 365 and 355 K, the interfacial thicknesses similarly increased with t at first and then turned to be invariant with respect to the annealing time, as shown in the parts (c) and (d) of Fig. 6.9. The evolved interfacial thickness at 365 and 355 K were 9.6 ± 2.5 and 11.4 ± 0.9 nm, respectively. Half of these values, namely surface mobile layer thicknesses, are much smaller than the unperturbed chain dimension, implying that segmental diffusion dominates the interfacial broadening of the (hPS/dPS) bilayers at 365 and 355 K rather than center-of-mass diffusion.

6.2.4 Summary

In this subsection, the features of SIMS measurements are briefly presented. SSIMS and DSIMS provide the top surface and slightly deeper region of soft materials.

References

1. Hirata T, Matsuno H, Tanaka M, Tanaka K (2011) Surface segregation of poly(2-methoxyethyl acrylate) in a mixture with poly(methyl methacrylate). PCCP 13(11):4928–4934. https://doi.org/10.1039/C0cp02101f
2. Vogler EA (1998) Structure and reactivity of water at biomaterial surfaces. Adv Colloid Interface Sci 74:69–117. https://doi.org/10.1016/s0001-8686(97)00040-7
3. Owens DK, Wendt RC (1969) Estimation of surface free energy of polymers. J Appl Polym Sci 13(8):1741–1747. https://doi.org/10.1002/app.1969.070130815
4. Sanada N, Yamamoto A, Oiwa R, Ohashi Y (2004) Extremely low sputtering degradation of polytetrafluoroethylene by C60 ion beam applied in XPS analysis. Surf Interface Anal 36(3):280–282. https://doi.org/10.1002/sia.1680
5. Ashley JC (1980) Inelestic interacitons of low-energy electrons with organic-solids—simple formulas for mean free paths and stopping powers. IEEE Trans Nucl Sci 27(6):1454–1458
6. Kawaguchi D, Tanaka K, Kajiyama T, Takahara A, Tasaki S (2003) Surface composition control via chain end segregation in blend films of polystyrene and poly(vinyl methyl ether). Macromolecules 36(18):6824–6830. https://doi.org/10.1021/ma034117u
7. Paynter RW (1981) Modification of the Beer-Lambert equation for application to concentration gradients. Surf Interface Anal 3(4):186–187. https://doi.org/10.1002/sia.740030410

8. Bhatia QS, Pan DH, Koberstein JT (1988) Preferential surface-adsorption in miscible blends of polystyrene and poly(vinyl methyl-ether). Macromolecules 21(7):2166–2175. https://doi.org/10.1021/ma00185a049
9. Schmidt I, Binder K (1985) Model-calculations for wetting transition in polymer mixtures. Journal De Physique 46(10):1631–1644. https://doi.org/10.1051/jphys:0198500460100163100
10. Van Vaeck L, Adriaens A, Gijbels R (1999) Static secondary ion mass spectrometry: (S-SIMS) Part 1. Methodology and structural interpretation. Mass Spectrom Rev 18(1):1–47
11. Tanaka K, Kajiyama T, Takahara A, Tasaki S (2002) A novel method to examine surface composition in mixtures of chemically identical two polymers with different molecular weights. Macromolecules 35(12):4702–4706. https://doi.org/10.1021/ma011960o
12. Eynde XV, Bertrand P, Jerome R (1997) Molecular weight effects on polystyrene fingerprint time-of-flight secondary ion mass spectrometry (ToF-SIMS) spectra. Macromolecules 30(21):6407–6416
13. Kawaguchi D, Tanaka K, Kajiyama T, Takahara A, Tasaki S (2003) Mobility gradient in surface region of monodisperse polystyrene films. Macromolecules 36(4):1235–1240. https://doi.org/10.1021/ma025667f
14. Kawaguchi D, Tanaka K, Takahara A, Kajiyama T (2001) Surface mobile layer of polystyrene film below bulk glass transition temperature. Macromolecules 34(18):6164–6166. https://doi.org/10.1021/ma010012k
15. Whitlow SJ, Wool RP (1991) Diffusion of polymers at interfaces—a secondary ion mass-spectroscopy study. Macromolecules 24(22):5926–5938. https://doi.org/10.1021/ma00022a006
16. Tanaka K, Takahara A, Kajiyama T (2000) Rheological analysis of surface relaxation process of monodisperse polystyrene films. Macromolecules 33(20):7588–7593. https://doi.org/10.1021/ma000406w

Chapter 7
Scattering and Reflection

Hideki Matsuoka

7.1 Static and Dynamic Light Scattering (SLS, DLS)

Static light scattering (SLS) is the technique to evaluate the size, weight-averaged molecular weight (Mw), and the second virial coefficient A_2 of the solutes in solution or dispersion by using laser light as an incident beam [1, 2]. Scattering intensity is measured as a function of the scattering angle (θ) (and concentration for A_2). Dynamic light scattering (DLS) is the technique to evaluate the hydrodynamic radius (R_h) by estimation of the translational diffusion coefficient D of the scattering particle by measuring time fluctuation of the scattering intensity [3].

When the size of the scattering particle is small enough compared to the wavelength of the incident laser beam, the reduced scattering intensity R_θ (the scattering intensity from the scatterer, i.e., the scattering intensity from the solution after subtracting the scattering from solvent only) is expressed as

$$R_\theta \equiv \frac{i_\theta r^2}{I_0(1+\cos\theta)} \tag{7.1}$$

where I_0 is the intensity of the incident light, i_θ is the scattering intensity at scattering angle θ, and r is the distance between the sample and detector.

The optical constant K is defined as follows:

$$K = \frac{2\pi^2 n_0^2\left(\frac{dn}{dc}\right)^2}{N_A\lambda_0^4} \tag{7.2}$$

H. Matsuoka (✉)
Department of Polymer Chemistry, Graduate School of Engineering, Kyoto University,
Kyoto-Daigaku Katsura, Nishikyo-ku, Kyoto 615-8510, Japan
e-mail: matsuoka.hideki.3s@kyoto-u.ac.jp

M. Maeda et al. (eds.), *Molecular Soft-Interface Science*,
https://doi.org/10.1007/978-4-431-56877-3_7

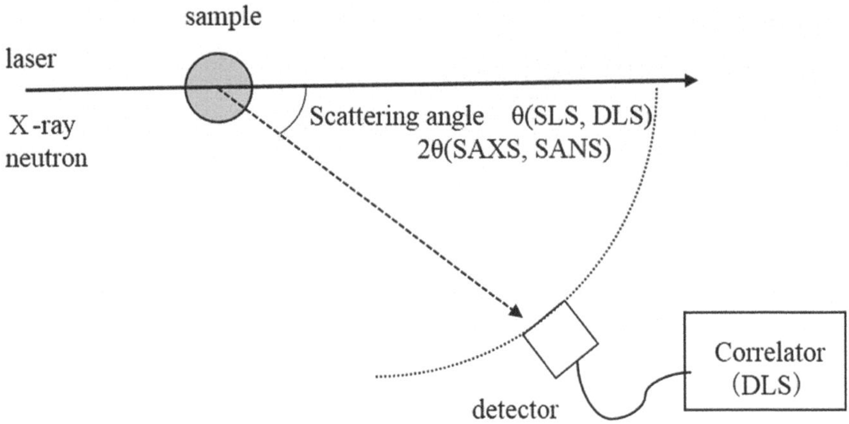

Fig. 7.1 Concept of scattering methods

where n_0 is the refractive index of the solvent, n is the refractive index of the solution, c is the concentration, λ_0 is the wavelength of the light in vacuum, and N_A is the Avogadro constant.

When θ is small enough, the following equation holds approximately:

$$\frac{Kc}{R_\theta} = \frac{1}{M_W}\left\{1 + \frac{\langle s^2\rangle}{3}\left(\frac{4\pi n \sin(\theta/2)}{\lambda}\right)^2\right\} + 2A_2c \tag{7.3}$$

where $<s^2>$ is the square average of the radius of the particle (Fig. 7.1).

Analysis of SLS is often carried out by following a so-called Zimm plot shown in Fig. 7.2 [1]. From the slope of the straight line which interpolated to $c = 0$, the radius of gyration can be estimated, and A_2 can be evaluated from the slope of the straight line which interpolated to $\theta = 0$. These two straight lines should have the same intercept for Y-axis, which is an inverse of Mw.

In DLS, the time fluctuation of scattered intensity as a function of time (in the order of ns - s) at fixed angles is collected, and calculate the time correlation function as schematically shown in Fig. 7.3. The intensity fluctuation directly reflects the motion (translational diffusion) of the scatterer in solution or dispersion. As shown in Fig. 7.3, for larger particle moving relatively slowly, the fluctuation is also slow, and the time correlation function shows slow decay. On the other hand, the smaller particle can move faster, so the short-time fluctuation is observed and the decay of the correlation function becomes faster. From this decay rate, the translational diffusion coefficient D can be estimated.

The correlator in DLS instrument gives the time correlation function of scattered intensity $g^{(2)}(q, \tau)$, and this gives us the time correlation function of the scattered field by the following equation called the Siegert relation:

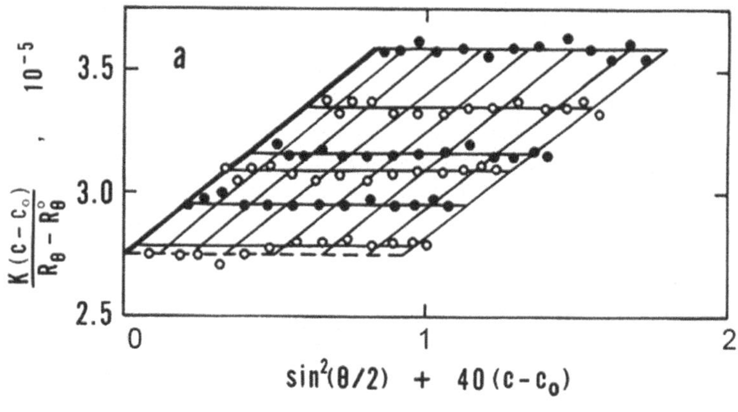

Fig. 7.2 Zimm plot of SLS data for hexadecyltrimethylammonium chloride micelle in aqueous solution with 0.5 M NaCl. From Imae and Ikeda [1]

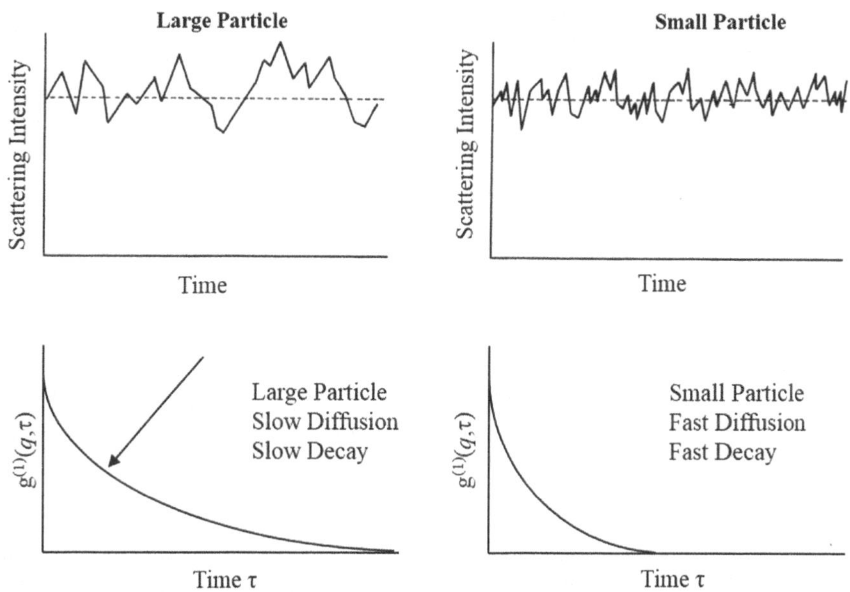

Fig. 7.3 (Top) Time fluctuation of the scattered light and (bottom) time correlation function for large (left) and small (right) particles

$$g^{(2)}(q,\tau) = A\left(1 + \left|g^{(1)}(q,\tau)\right|^2\right) \tag{7.4}$$

where q is the scattering vector defined with the scattering angle θ by

$$q = 4\pi \sin(\theta/2)/\lambda \tag{7.5}$$

and τ the decay time, λ the wavelength of the light and A the baseline.

For Brownian particles,

$$g^{(1)}(q,\tau) = \exp(-\Gamma\tau) \tag{7.6}$$

and the decay rate Γ is expressed as

$$\Gamma = Dq^2 \tag{7.7}$$

Hence, the translational diffusion coefficient D can be evaluated by the slope of the experimentally obtained Γ versus q^2 plot, after measurements at several scattering angle θ. By assuming the Stokes–Einstein equation, the hydrodynamic radius of the particle can be evaluated.

$$R_h = \frac{k_B T}{6\pi\eta D} \tag{7.8}$$

where k_B is the Boltzmann constant, T is the absolute temperature, and η is the viscosity of the solvent.

Figure 7.4 shows an example of $g^{(1)}(q, \tau)$ and Γ versus q^2 plot for polymer micelle in aqueous solution [4]. In this case, since a small amount of large aggregate is included in the solution, the so-called double-exponential fitting was employed and two R_h values, for polymer micelles and aggregates, were obtained by the fitting shown in solid line in Fig. 7.4a. An excellent linearity in Γ versus q^2 plot (Fig. 7.4b) with passing through the origin guarantees that the time correlation function obtained certainly reflects the intensity fluctuation by the translational diffusion. If the time correlation function includes the contribution of the rotational diffusion, this plot does not go through the origin, and if influenced by the internal motion of the particle, this plot does not become straight line.

R_h evaluation is limited to a very diluted solution or dispersion without interparticle interaction. For higher concentration systems, in which interparticle interaction exists, the motion influenced by interaction and hydrodynamic size of the "cluster" can be estimated if it exists. Figure 7.5 shows $g^{(1)}(q, \tau)$ for poly(styrenesulfonate) (PSS), an polyelectrolyte, in aqueous solution is shown [5]. Obviously, two decay modes are observed. One is the motion of single PSS influenced by interaction, and the other, slower is the diffusion of PSS cluster. It might be interesting to note that both of these dynamic modes are influenced by counterion species. Na^+ and H^+ as a counterion show different decay rates.

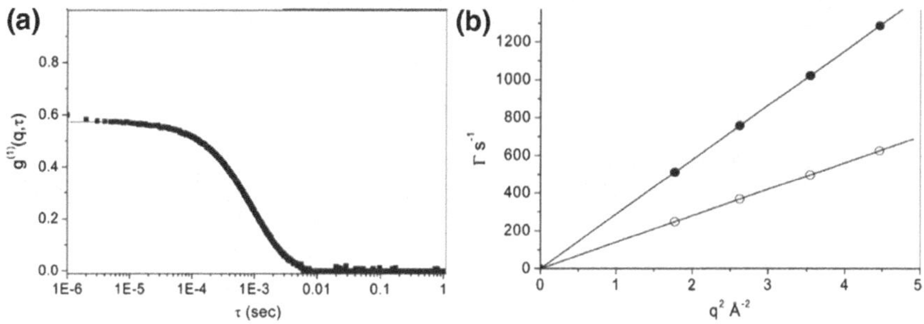

Fig. 7.4 **a** The time correlation function of scattered field $g^{(1)}(q, t)$ and **b** G versus q^2 plot for cationic amphiphilic diblock copolymer micelles in aqueous solution obtained by DLS. [Polymer] = 0.2 mg/ml. The time correlation function was fitted by double-exponential function (solid line in (**a**)), and the fast mode is diffusion of micelle and the slow mode is for large aggregate in solution. The hydrodynamic radius (R_h) of the micelle was estimated to be 85 nm from the slope of the straight line for the fast mode in (**b**). From Ghosh et al. [4]

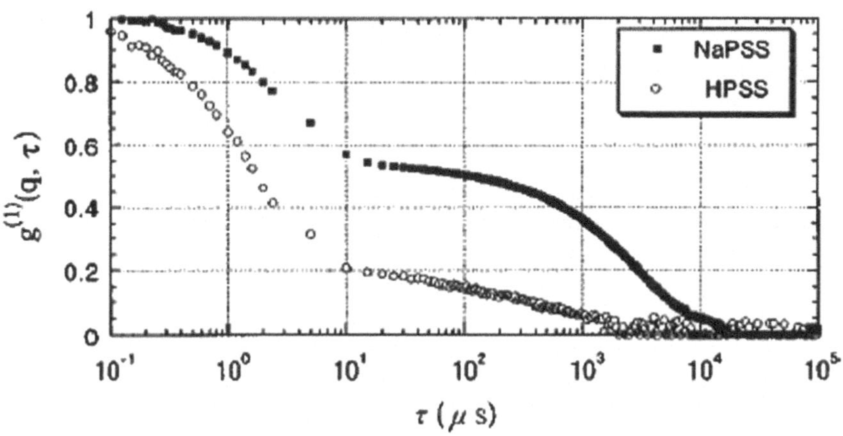

Fig. 7.5 Time correlation function of scattered field for poly(styrenesulfonate) aqueous solutions. Two dynamic modes are observed, which correspond to polymer single molecule diffusion (fast) and polymer cluster diffusion (slow). Both are influenced by counterion species (Na^+ or H^+). [Polymer] = 0.05 monomer mole/liter. From Matsuoka et al. [5]

ζ-potential is an important factor characterizing the situation of surface charge of colloidal particles. There are some techniques to evaluate ζ-potential, but the electrophoretic light scattering (ELS) is one of the most convenient ones. In principle, ζ-potential is calculated from electrophoretic mobility, which was evaluated by Doppler shift in ELS experiment. For the details, refer to textbooks [3].

Total internal reflection microscopy (TIRM) or evanescent wave light scattering microscopy (EVLSM) is a powerful tool developed by D.C. Prieve to evaluate particle–surface interaction potential [6]. A large negatively charged colloidal particle sediments down to the bottom glass surface, but if this surface is negatively charged,

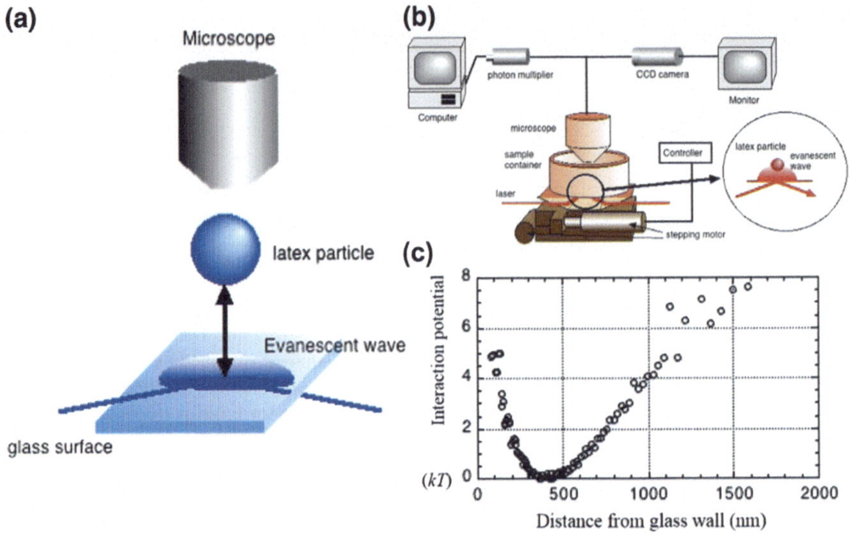

Fig. 7.6 **a** The principle and **b** experimental setup of TIRM. **c** The interaction potential between latex particle and charged glass wall evaluated by TIRM. From Matsuoka [7, 8]

by the electrostatic repulsion between the surface and particle, the particle repeats up and down motions by thermal energy in the potential well created by superposition of gravity and this electrostatic repulsion. If the particle in this situation is irradiated by evanescent wave, the distance between the particle and the bottom glass surface can be estimated by the scattering intensity of evanescent wave since the evanescent wave intensity decays exponentially as a function of the distance from the glass surface. If we measure this distance repeatedly, the interaction potential curve can be evaluated as a frequency distribution of the particle position (Fig. 7.6) [7, 8]. While the atomic force microscopy (AFM) can measure the "force" between the particle and wall, TIRM gives us the particle-wall intercalation "potential" directly, which might be a unique technique.

TIRM is, in principle, static measurement, but evanescent wave dynamic light scattering (EVDLS) is also possible, if we measure the time fluctuation of intensity of scattered evanescent wave [8, 9]. In this case, the diffusion coefficient of the colloidal particle interacting with glass wall can be estimated as a function of the distance between the particle and glass surface. One example is shown in Fig. 7.7. Obviously, the diffusion coefficient is small (slow diffusion) when the particle is near the wall, but it increases with increasing the distance and finally becomes equal to that for free Brownian particle. Application of evanescent wave to scattering technique should attract attention as a new tool for colloidal and surface studies.

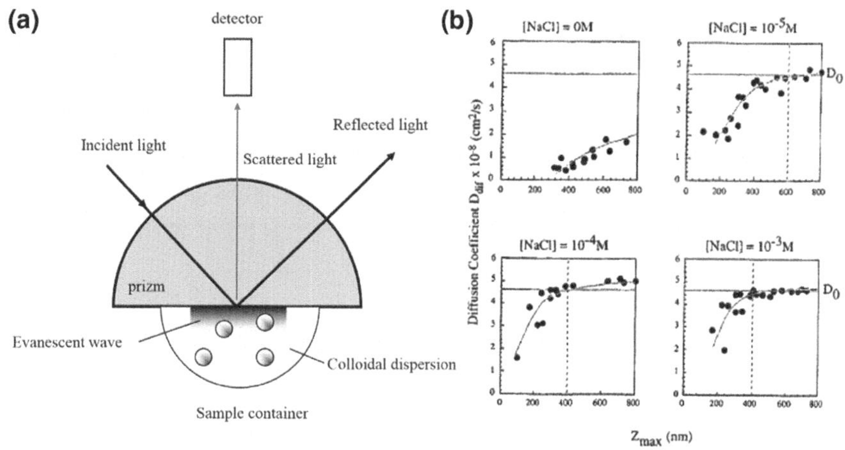

Fig. 7.7 **a** Principle of EVDLS and **b** diffusion coefficient of latex particles near glass wall as a function of the distance from the wall surface. From Matsuoka et al. [8, 9]

7.2 Small-Angle X-Ray and Neutron Scattering (SAXS, SANS)

Small-angle X-ray scattering (SAXS) and small-angle neutron scattering (SANS) are very powerful techniques for structure analysis in the scale of 15Å–1000Å (Glatter and Kratky [10]. The scattering intensity (I) is proportional to the number of scattering object (N), to square of its volume (V), and to square of density difference ($\Delta\rho$) between inside and outside of the scatterer (contrast).

$$I \propto NV^2(\Delta\rho)^2 \tag{7.9}$$

The density difference, $\Delta\rho$, is an electron density difference for X-ray, scattering length density difference for neutron, respectively.

As shown in Fig. 7.8, scattered light from the large particle is enhanced in a smaller angle region, the scattering intensity below, typically, 5° is measured for polymer, micelle, and colloid studies, which is the reason why this technique is called "small-angle scattering". For the smaller particle, scattering at larger scattering angle (2θ) can be detected as curve 1 in Fig. 7.8 while not for larger particles (curve 2). For an anisotropic particle, such as a cylinder, which has both large and small character for length and diameter, respectively, the scattering curve is like curve 3, in which both of large and small characters are superimposed.

One scattering curve, i.e., the scattering intensity versus scattering angle plot, contains information on the nanostructure of scattering particles. Figure 7.9 shows the information in the scattering curve schematically. In the analysis of scattering data, the scattering vector, q, is used instead of the scattering angle 2θ, defined as

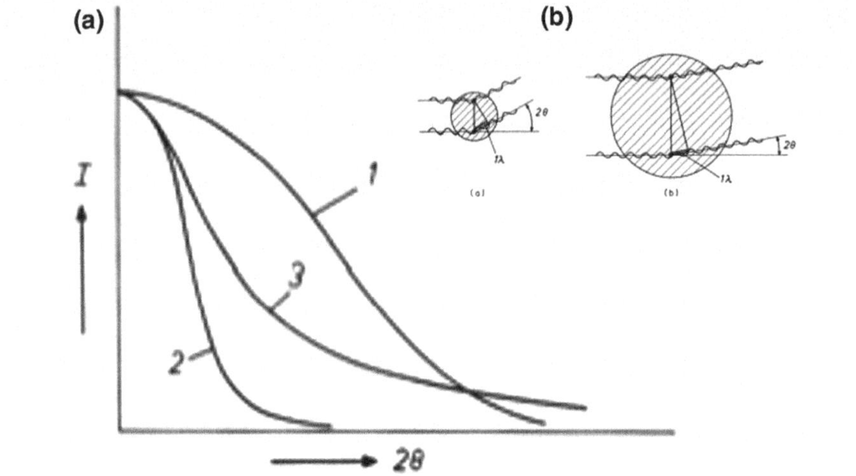

Fig. 7.8 **a** Schematic representation of scattering curve for (1) small particle, (2) large particle, and (3) anisotropic particle; **b** scattering geometry for small and large particles. From Glatter and Kratky [10]

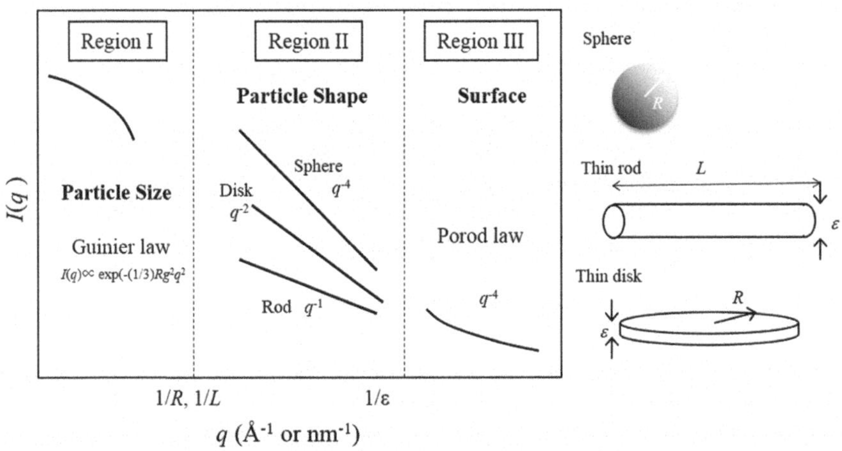

Fig. 7.9 Information obtained by small-angle scattering

$$q = 4\pi \sin(\theta)/\lambda \tag{7.10}$$

where λ is the wavelength of the incident beam. The position of Bragg peak for certain lattice spacing appears at different scattering angles if the wavelength of the incident light is different, but it appears at the same q position independent of the wavelength the incident beam.

Since the dimension of q is the inverse of length (Å^{-1}, or nm^{-1}), the scattering curve at smaller q region contains structure information in large scale, while that at larger q in smaller scale. The q region is often divided into three regions as is in Fig. 7.8. The region I at smaller q gives us an average size of the scattering particle as the radius of gyration, Rg, since we are looking at the particle with large scale.

In region I, the scattering follows the Guinier law:

$$I(q) \propto e^{-\frac{1}{3}Rg^2q^2} \quad (7.11)$$

Hence, the initial slope of the so-called the Guinier plot, i.e., ln $I(q)$ versus q^2, the radius of gyration of the scattering particle can be evaluated. For a homogeneous sphere of radius R,

$$Rg = \sqrt{\frac{3}{5}}R \quad (7.12)$$

and for infinitely thin hollow sphere of radius R, then

$$Rg = R \quad (7.13)$$

holds. The relationship between the geometrical size and Rg can be found in the fundamental textbook [10].

In the middle region II, we are looking at the scattering particles in the same scale of particle size, the information about the shape, such as sphere, rod, disk, can be elucidated. In the region III at larger q region, we are looking at the particles with smaller scale than particle size, we can get no information of the size and shape of the particles anymore, but the information of the surface of the particle can be evaluated. If $I(q)$ in this region is proportional to q^{-4}, this means that the surface of the particle is perfectly smooth, which is called the Porod law. If the surface is not smooth but there is some geometrical roughness on the surface, $I(q)$ is proportional to the power between -4 and -3 depending on roughness. In this case, the surface Fractal dimension, D, is evaluated by $D = 6–\alpha$, where -α is the power of $I(q)$ to q. The border of each region depends on the size of the scatterer as shown in Fig. 7.9.

Figure 7.10 shows examples of SANS curves for ionic amphiphilic diblock copolymer micelles in aqueous solution with and without added NaCl (Kawesaiha et al. [11]). Since the micelle has usually internal structure such as core-shell structure, the scattering curves are often analyzed by model fitting. The fitting curve for spherical core-shell model is calculated with the density of the core, radius of the hydrophobic core, shell thickness as fitting parameters. The density of the core is often equal to that of hydrophobic block, so the aggregation number N_{agg} of the micelle can be calculated by the volume of the core if the partial molar volume of the hydrophobic block is known. The density of the corona is not a fitting parameter, since it can be calculated by volume fraction of hydrophilic block chain (its number in the shell should be equal to N_{agg}) and solvent (water). In this model, the constant density in the shell is assumed although the "shell" is corona consists of hydrophilic chains and solvent, i.e., not homogeneous. Hence, the fitting curve by core-shell

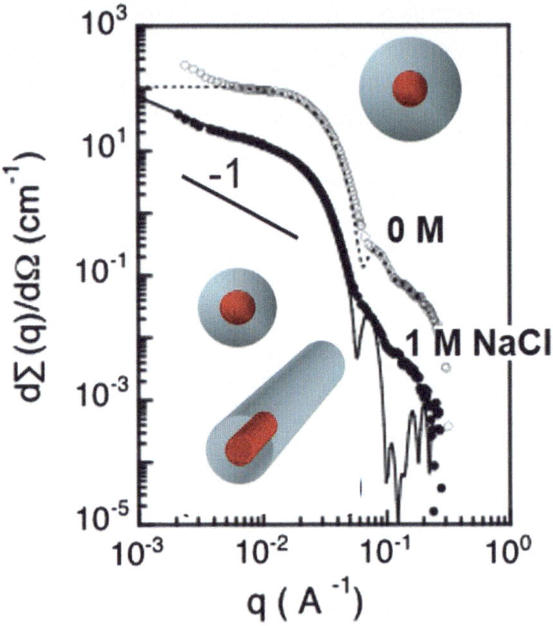

Fig. 7.10 SANS profiles for ionic amphiphilic diblock copolymer micelles in aqueous solutions without and with 1 M NaCl. The profile for no salt system is well fitted by Pedersen model, from which core radius = 60Å, shell thickness = 44Å, and aggregation number = 108 are obtained. The profile for 1 M NaCl system is well fitted by model of mixture of spherical and cylindrical micelles with same core radius and shell thickness. The volume fraction of spherical micelle was estimated to be 25%. The y-axis is absolute intensity of scattered neutron. From Kaewsaiha et al. [11]

model cannot give a good agreement at higher q regions for polymer micelle. In such a case, the Pedersen model is used, in which the shell is assumed to consist of Gaussian coil of hydrophilic polymer chains and solvent. For the details of this model, refer to original articles [12]. The fitting curve for no salt system indicated by the dotted line in Fig. 7.10 is based on the Pedersen model. Very good agreement between experimental and fitting curves is obtained. A small upturn at small-angle region in the experimental curve is due to large aggregates, so it is neglected in this fitting procedure.

Ionic polymer micelles transformed to rod-like micelles as conventional ionic low molecular weight surfactant by salt addition. This is simply due to the request of the critical packing parameter proposed by Islaerativilli [13]. In this process, it is not the case that all the spherical micelles change to be rod-like at certain salt concentration, but spherical and rod-like micelles coexist and fraction of rod-like micelle increases with increasing added salt concentration. The fitting curve in Fig. 7.10 shown by the solid line for 1 M NaCl system is calculated assuming a mixture of core-shell spherical micelle and core-shell rod-like micelle having the same core radius and shell thickness. Very good agreement was obtained and the volume fraction of spherical and rod-like micelles in the mixture can be elucidated.

Bonse and Hart proposed a special small-angle scattering instrument in 1966 [14]. It has a very special optical alignment by using two channel-cut single crystals (Si or Ge), which is largely different from that shown in Fig. 7.1. Refer to the original articles for the details and principles of this small-angle technique, but this is often called ultrasmall-angle scattering (USAXS) since it can detect the scattering intensity at very small-angle regions, which is impossible for normal scattering instruments due to the contribution of direct beam. This very high resolution at very small-angle region enables us to apply this technique for the structure study in the range of 1000Å–30000Å (3 μm). Hence, this technique is very powerful for the study of concentrated colloidal dispersions, which cannot be investigated by light scattering due to is high turbidity.

For dilute solutions and dispersions, $I(q)$ reflects only nanostructure of scattering particle. However, with increasing concentration, interaction (steric and/or electrostatic) between scattering particles influences the spatial distribution of particles in solution or dispersion. In such a case, interparticle interference is included in the scattering curve. For spherical systems,

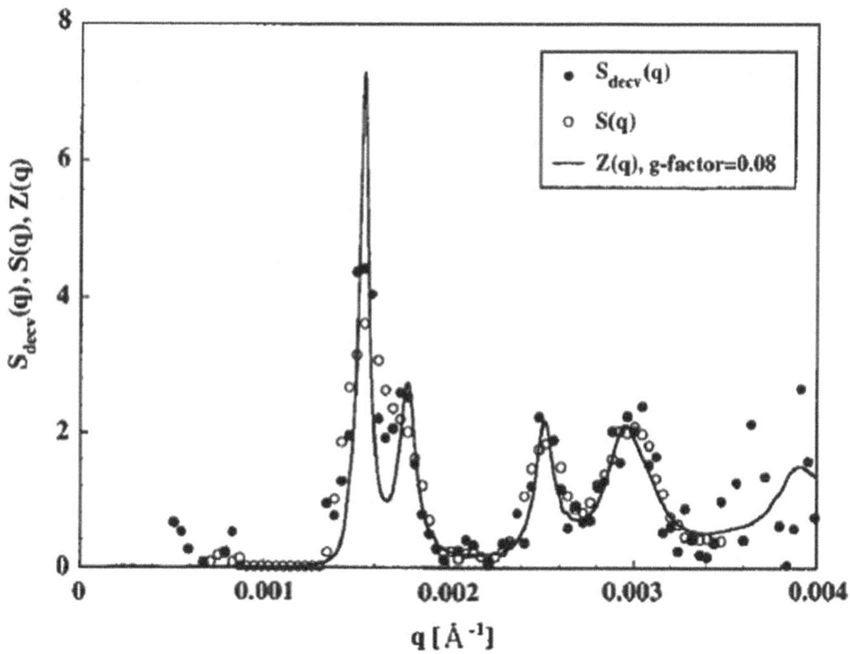

Fig. 7.11 The structure factor $S(q)$ for colloidal crystal in latex dispersion obtained by USAXS. The relative position of Bragg peaks indicates fcc structure of the crystal. Solid line is the fitting by 3D-paracrystal lattice factor $Z(q)$. The degree of distortion, g-factor, was estimated to be 0.08 by this fitting. From Harada et al. [15]

$$I(q) = P(q)\,S(q) \tag{7.14}$$

where $P(q)$ is the particle scattering factor and $S(q)$ is the structure factor, which reflects the spatial distribution of center of the mass of the spherical particles. For diluted solution, $S(q) = 1$ for any q, we can evaluate the size and shape of the particle from $I(q)$ because $I(q) = P(q)$. If we know $P(q)$ in advance, we can extract $S(q)$ function by dividing $I(q)$ by $P(q)$, which provide us higher order arrangement of the particle in the solution or dispersion.

Figure 7.11 shows the $S(q)$ curve for colloidal crystal in dispersion obtained by USAXS and its fitting results based on 3D-paracrystal theory [15]. The position of Bragg peaks tells us that this colloidal crystal is the face-centered cubic lattice. By the fitting, the degree of distortion of the crystal (g-factor) can be estimated to be 0.08. The g-factor is 0 for perfect crystal, and it increases with an increase of distortion (the distortion of the second kind). By the concept of Ziman, ca.$g = 0.15$ ($S(q) = 2.85$) is the transition point between solid crystal and liquid, i.e., melting point. As shown by this example, the small-angle technique can be applied not only to particle size/shape studies but also solution/dispersion structural studies.

References

1. Imae T, Ikeda S (1989) Characteristics of rodlike micelles of Alkyltrimethylammonium halides in aqueous sodium halide solutions: their flexibility and entanglement, surfactants in solution **7**, 455
2. Chu B (1991) Laser light scattering, 2nd ed., Academic Press
3. Pecora R ed. (1985) Dynamic light scattering, plenum
4. Ghosh A, Yusa S, Matsuoka H, Saruwatari Y (2011) Non-surface activity and micellization behaviour of cationic amphiphilic block copolymer synthesized by reversible addition—fragmentation chain transfer process. Langmuir 27(15):9237–9244
5. Matsuoka H, Ogura Y, Yamaoka H (1998) Effects of counterioin species on the dynamics of polystyrenesulfonate in aqeous solution as studied by dynamic light scattering. J Chem Phys 109:6125
6. Prieve DC, Frej NA (1990) Total internal reflection microscopy: a quantitative tool for the measurement of colloidal forces. Langmuir 6:396–403
7. Tanimoto S, Matsuoka H, Yamaoka H (1995) Direct evaluation of dynamic characteristics and interaction potential between colloidal particle and glass wall by evanescent wave light scattering microscope method. Colloid Polymer Sci 273:1201
8. Matsuoka H (2001) Evanescent wave light scattering—fusion of evanescent wave and light scattering techniques to the study of colloids and polymers near interface. Macromol Rapid Commun 22(2):51
9. Matsuoka H, Morikawa H, Tanimoto S, Kubota A, Naito Y, Yamaoka H (1998) Evaluation of dynamic property of polymer latex particles interacting with quartz interface by evanescent wave dynamic light scattering. Colloid Polym Sci 276(3):349–355
10. Glatter O, Kratky O (1982) Small angle X-ray scattering, Academic
11. Kaewsaiha P, Matsumoto K, Matsuoka H (2007) Sphere to rod transition of non-surface active amphiphilic Diblock copolymer micelles—a small-angle neutron scattering study. Langmuir 23(18):9162–9169
12. Pedersen JS, Posselt D, Mortensen K (1990) Analytical treatment of the resolution function for small-angle scattering. J Appl Crystallogr 23, 321

13. Israelachvili J (1991) Intermolecular and surface forces, 2nd ed., Academic
14. Bonse U, Hart M (1966) Analytical treatment of the resolution function for small-angle scattering. Z Physik 189:151
15. Harada T, Matsuoka H, Yamaoka H (1999) An exact evaluation of salt concentration dependence of interparticle distance in colloidal crystals by ultra-small-angle X-ray scattering.III - confirmation of solid-liquid transition by 3D-Paracrystal analysis. Langmuir **15**(2), 573
16. Matsuoka H, Morikawa H, Yamaoka H (1996) Rotational diffusion of ellipsoidal latex particles in dispersion as studied by depolarized dynamic light scattering. Colloids Surf A 109:137–145

Chapter 8
X-Ray and Neutron Reflectivity and Grazing Incidence X-Ray Diffraction

Atsushi Takahara and Yuji Higaki

8.1 Background

Ordered structure formation is the essential concept to control the soft materials and the basics of bottom-up molecular technology. Various well-defined molecular assemblies are found in the nature such as lipid bilayer and DNA hybridization. The well-defined supramolecules have been investigated with inspirations from the nature. In order to fabricate functional soft materials, the control of nanostructure and hierarchical assembly are required.

The scattering (or reflection) of X-rays and neutrons from surfaces offers valuable insights for the nature of surfaces and interfaces, and the methods complement to other techniques for surface structure study such as scanning force microscopy (SFM), secondary ion mass spectroscopy (SIMS), X-ray photoelectron spectroscopy (XPS), sum frequency generation (SFG) spectroscopy, etc. The technique of X-ray and neutron reflectivity measurement has undergone rapid development in recent years, with expanding applications.

The intensity of scattering in the small-angle regime obeys the Porod law as the scattering vector $Q\left(= 4\pi \frac{\sin\theta}{\lambda}\right)$ increase, and the absolute value of the intensity is proportional to the total area of interfaces within the sample. It was also shown that the deviation of the observed intensity curve from the Porod law can give the diffuseness of the interfaces. The reflectivity measurement is regarded as an extension of the Porod law to surfaces that are essentially flat, and interfaces that are close to exposed surfaces and parallel to them. As small-angle scattering, we obtain scatterings at

A. Takahara (✉)
Institute for Materials Chemistry and Engineering, Graduate School of Engineering, Kyushu University, 744 Motooka, Nishi-ku, Fukuoka 819-0395, Japan
e-mail: takahara@cstf.kyushu-u.ac.jp

Y. Higaki
Department of Integrated Science and Technology, Faculty of Science and Technology, Oita University, 700 Dannoharu, Oita 870-1192, Japan
e-mail: y-higaki@oita-u.ac.jp

M. Maeda et al. (eds.), *Molecular Soft-Interface Science*,
https://doi.org/10.1007/978-4-431-56877-3_8

small Q, in which the radiation strikes the surface at a small grazing angle and is reflected from the surface at a similarly small angle.

When an X-ray beam impinges on the surface of a sample, similar considerations apply as for optics with visible light, i.e., the refractive index n is the most important parameter. For X-rays, i.e., electromagnetic radiation with a wavelength λ around 0.1 nm, the refractive index is defined as

$$n = 1 - \delta + i\beta$$

with

$$\delta = \frac{\lambda^2}{2\pi} r_e \rho_e$$

and

$$\beta = \frac{\lambda}{4\pi} \mu_x$$

where $r_e = 2.818 \times 10^{-15}$ m is the classical electron radius, ρ_e is the electron density of the material, and μ_x is the absorption length. With $\delta > 0$, we find that $n < 1$, which leads to the phenomenon of total external reflection for incident angles α below the critical angle $\alpha_c \simeq \sqrt{2\delta} = \lambda\sqrt{r_e \rho_e / \pi}$. Typical values for δ are 10^{-5}–10^{-6}, and thus α_c is in the range of 0.1°–0.5°. For simplicity, β, which is even smaller, will be ignored here [1, 2].

An analogous description of the index of refraction applies to the case of neutrons, as described by

$$\delta = \frac{\lambda^2}{2\pi} b\rho_n \text{ and } \beta = \frac{\lambda}{4\pi} \mu_n$$

The order of magnitude for δ and β is similar to the case of X-ray. However, since the scattering length b of the nuclei varies non-monotonously across the periodic table, as opposed to the case of X-rays, the contrast between two given elements is different for X- rays and for neutrons. Thus, X-ray and neutron reflectivity can be used in a complementary fashion. Moreover, neutrons offer a contrast even between different isotopes. The most famous example is the large difference in b between H and D, i.e., hydrogen with protons or deuterons as nuclei. Since hydrogen is ubiquitous in organic matters, deuteration is a frequently applied method to obtain a contrast between different organic materials [2].

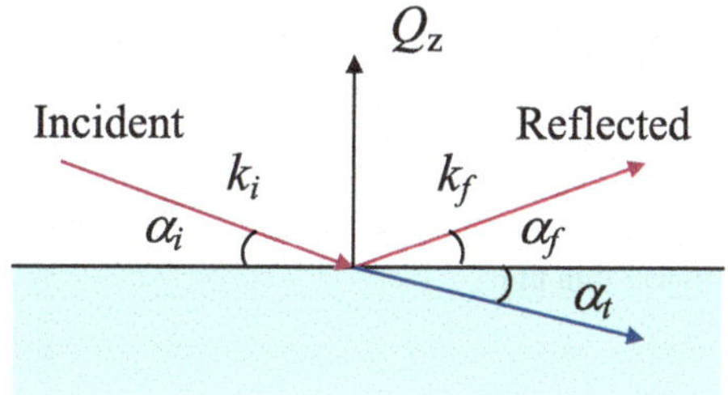

Fig. 8.1 Schematic representation of geometry for specular reflectivity. The incident and the reflected wave vectors, k_i and k_f, define the scattering plane

8.2 Reflectivity

In this section, we only concerned with the specular part, i.e., the incident angle α_i and exit angle α_f are equal. In this case, the momentum transfer $\vec{Q} = \vec{k}_f - \vec{k}_i$ is surface normal, which is chosen to be the z coordinate. For a given wavelength λ, momentum transfer is derived to be (Fig. 8.1)

$$Q_z = 4\pi \frac{\sin\alpha_i}{\lambda}$$

The complex reflection coefficient of the electrical field for an ideal, sharp interface is described by Fresnel coefficient

$$r(Q) = \frac{k_z - k_z'}{k_z + k_z'} = \frac{Q - Q_t}{Q + Q_t}$$

where k_z (=Q) and k'_z (=Q_t) are the vertical component of the incident and transmitted waves, respectively. The refracted wave vector can be defined as

$$Q_t^2 = Q^2 - Q_c^2$$

where Q_c is the wave-vector transfer at the critical angle α_c (Q_c). Then,

$$r(Q) = \frac{Q - \sqrt{Q^2 - Q_c^2}}{Q + \sqrt{Q^2 - Q_c^2}} = \frac{\sin\alpha_i - \sqrt{\sin^2\alpha_i - \sin^2\alpha_c - 2i\beta}}{\sin\alpha_i + \sqrt{\sin^2\alpha_i - \sin^2\alpha_c - 2i\beta}}$$

The reflected intensity, R_F, is expressed as

$$R_F = rr^* = |r|^2$$

When the wave-vector transfer Q is far large to Q_c, the reflected intensity is described by the following asymptotic equation (Fig. 8.2):

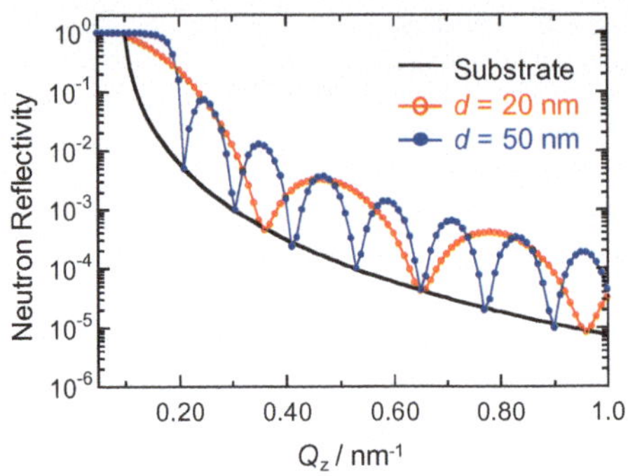

Fig. 8.2 Calculated specular Fresnel reflectivity from a flat Si surface (black line) and reflectivity in the presence of a 20 nm (red plots) or a 50 nm (blue plots) polymer thin film

$$R_F = \left(\frac{Q_c}{2Q}\right)^4$$

Above the critical angle, where k_z' is not very different from k_z, the reflected intensity falls off rapidly ($R_F \propto 1/\alpha_i^4$) following Porod law. Surface roughness σ effect on reflectivity can be expressed by

$$r_{rough} = r_{ideal} exp\left(-\frac{1}{2} Q Q_t \sigma^2\right)$$

If the sample has more than one interface like a case of a polymer thin film on a substrate, the scattering from all interfaces has to be considered. Parratt developed a formalism, which relates the reflected and transmitted amplitude, R_j and T_j, respectively, by following equations [3]:

$$X_j = \frac{R_j}{T_j} = e^{-2ik_{zj}z_j} \frac{r_{j,j+1} + X_j e^{2ik_z z_j}}{1 + r_{j,j+1} X_{j+1} e^{2ik_z z_j}}$$

where

$$r_{j,j+1} = \frac{k_{z,j} - k_{z,j+1}}{k_{z,j} + k_{z,j+1}}$$

is the Fresnel coefficient of interface j. The recursion is solved using $T_1 = 1$ and $R_{N+1} = 0$ (no reflection from the substrate).

Due to the interference of waves, which are reflected from different interfaces within a system, intensity oscillations in the reflectivity are observed. The periodicity in Q_z (Kiessig fringes) is associated with the thickness t of the film by the following equation:

$$t \approx \lambda/2\Delta Q$$

where ΔQ is the interval between successive maxima or minima. The roughness of the interfaces is taken into account in the same fashion by including a term $e^{-2k_{z,j},k_{z,j+1}\sigma_{j,j+1}^2}$ in the Fresnel coefficients if the roughness is small compared to thicknesses of the layers involved.

When we measure reflectivity, software such as MOTOFIT is used to evaluate electron density or scattering length density (SLD) profile. MOTOFIT is a package that aids the least squares fitting of specular X-ray and neutron reflectivity data, using an Abeles matrix formalism or Parratt's recursion formula [4].

8.3 Example of Reflectivity from Thin Films

Figure 8.3 shows the X-ray reflectivity profile of an amorphous polystyrene film at room temperature [5]. The XR data was acquired in SPring-8 BL03XU beamline using ultrahigh photon flux synchrotron radiation source produced by an undulator. The X-ray beam was precisely collimated to yield an excellent angular resolution. The XR profile was analyzed by fitting a single layer model on an Si-wafer with a native oxide layer. A calculated reflectivity based on the single layer model (solid line) gave good agreement with the reflectivity data. The thickness of the film was determined to be 34.8 nm and a surface roughness of 0.3 nm through the curve fitting analysis.

Figure 8.4a shows the neutron reflectivity (NR) profile of a deuterated polystyrene (d-PS) spin-cast thin film on a Si substrate (30 × 30 mm^2) at an angular resolution $\Delta\theta/\theta$ of 3.0% [6]. The solid line in the figure represents a best-fit curve based on model calculations assuming a three-layer model of d-PS/SiO_2/Si (Fig. 8.4b). The MOTOFIT program was used to fit the reflectivity profile to the model SLD layers, wherein the thickness of each layer, SLD, and Gaussian roughness were optimized to minimize the χ^2 between the measured and calculated reflectivity curves. The NR of d-PS film shows a total reflection at low Q ranges and several fringes between Q = 0.2–1.2 nm^{-1}. The observable reflectivity decreased by seven orders of magnitude over Q range of $0 < Q < 4$ nm^{-1}. The critical scattering vector (Q_c) of the d-PS film is 0.18 nm^{-1}, which is consistent with theoretical Q values calculated by SLD of d-PS (6.46 × 10^{-4} nm^{-2}), SiO_2 (3.47 × 10^{-4} nm^{-2}), and Si (2.07 × 10^{-4} nm^{-2}). The NR curve exhibits regular oscillation, so-called Kiessig fringes, with a constant period corresponding to the film thickness. The slope of the NR curve gently changes at Q = 1.0 nm^{-1}, which is attributed to a very thin layer of SiO_2 at the interface of d-PS and Si substrate. The NR curve is unable to reproduce without the thin SiO_2 layer at high Q region around 1–4 nm^{-1}. The fitting curve (solid line) in Fig. 8.4a reproduced NR plots very well up to $Q \sim 2$ nm^{-1} by applying the three-box model of d-PS (97.8 nm), SiO_2 (1.25 nm), and Si substrate. The thickness of d-PS estimated by NR was close to the thickness independently determined by spectroscopic ellipsometry (98.2 nm).

Figure 8.5a, b shows NR and XR curves of the cadmium (Cd)/deuterated stearate (d-SA) LB film as a function of Q, respectively. In the NR profile, four distinctive

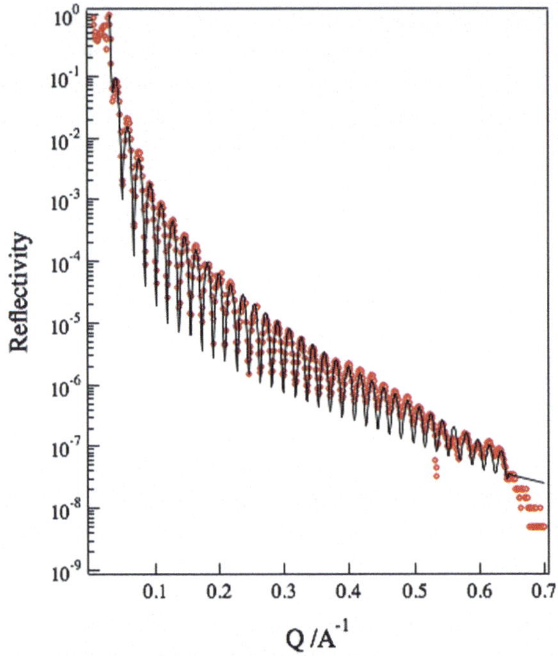

Fig. 8.3 X-ray reflectivity profile of a polystyrene thin film on a silicon wafer (dots). The solid curve is the fitting curve based on the Parratt algorithm. Reproduced with permission from [5]. Copyright (2013) Nature Publishing Group

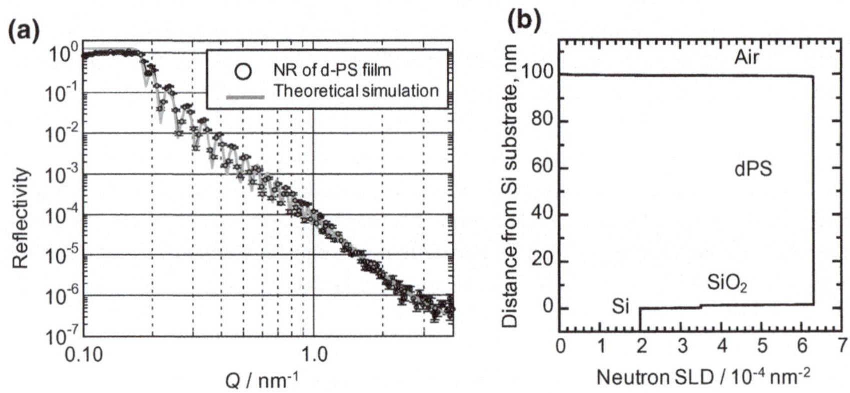

Fig. 8.4 **a** Neutron reflectivity of a deuterated polystyrene thin film, and **b** the corresponding SLD profile obtained by the curve fitting. Reproduced with permission from [6]. Copyright (2013) Nature Publishing Group

Bragg peaks were observed at $Q = 1.20, 2.46, 3.68,$ and $4.91\ \text{nm}^{-1}$, although the peak at $Q = 1.20\ \text{nm}^{-1}$ was not as clear as the corresponding peak in the XR profile, because the SLD contrast between Cd ions and alkyl groups was lower than the contrast in the electron density. These Bragg peak positions obtained from NR are in good agreement with those of XR, within ±5%, which is comparable to the angular resolution of 3.0%. The spacing of bilayers estimated from the Bragg peaks was 5.1 nm, which was smaller than the bilayer sequence of non-tilted stearic acid (5.5 nm), indicating that the d-SA alkyl groups slightly tilt and orient almost perpendicular to the film surface. From the Kiessig fringe periodicity, the total film thickness was determined to be 39 nm. Using these parameters and the MOTOFIT program, we estimated the neutron SLD and electron density profiles of 7.5 bilayers from the NR and XR curves (Fig. 8.5c, d). The obtained SLD value of d-SA varied from 5.0 to $5.9 \times 10^{-4}\ \text{nm}^{-2}$ inside the film except for the outermost 0.5 bilayer ($4.1 \times 10^{-4}\ \text{nm}^{-2}$), although the theoretical SLD value of d-SA is $5.8 \times 10^{-4}\ \text{nm}^{-2}$ as calculated using a mass

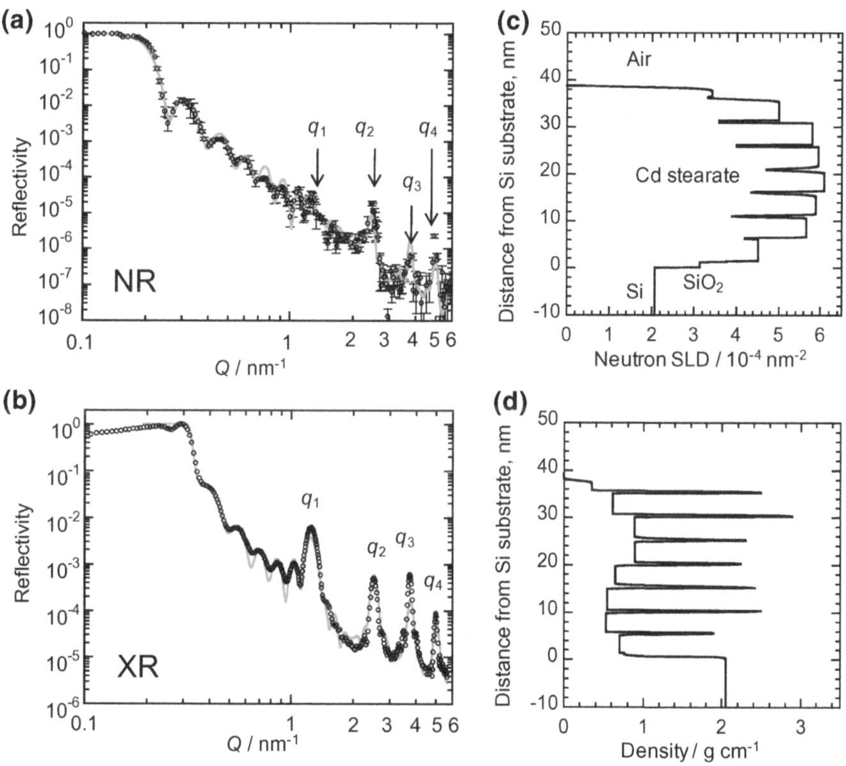

Fig. 8.5 **a** Neutron reflectivity and **b** X-ray reflectivity curves as a function of Q, and the corresponding **c** neutron SLD and **d** electron density profiles of 7.5 bilayers of cadmium deuterated stearate prepared on Si substrate by Langmuir–Blodgett method. Reproduced with permission from [6]

density of 0.84 g cm^{-3}. The SLD of the first bilayer in contact with the Si substrate was relatively low in comparison with that of bilayers 2–7. The electron density of the cadmium ion layer near the Si substrate was also lower than that of the upper layers, probably due to the difference in molecular aggregation structure.

NR measurement unravels water-soluble polymer brush structure under aqueous solutions. A water-soluble poly[{2-(methacryloyloxy)ethyl}trimethylammonium chloride] (PMTAC) brush film with M_n = 110,000 and relatively narrow molecular weight distribution (M_w/M_n = 1.19) was prepared by surface-initiated atom transfer radical polymerization (ATRP) from a silicon disk with a 3-inch diameter and 10 mm thickness. Although the PMTAC is hydrogenated polymer, structural analysis in polymer/deuterated solvent interfaces is possible due to the great SLD contrast. Figure 8.6 shows NR curves of the PMTAC brush in air and D_2O, and the corresponding neutron SLD profiles along with the distance from the Si disk surface. The thickness of PMTAC brush film under air at 298 K under 50% relative humidity was determined to be 29.0 nm from the periodic fringes using a five-layer model (air/PMTAC brush/surface initiator layer/SiO_2/Si).

On the other hand, the NR curve of the PMTAC brush in D_2O showed indistinguishable fringes; however, the specular reflection was measured down to 10^{-6} in reflectivity up to 2.0 nm^{-1} using white neutrons with wavelengths of 0.20–0.88 nm. The SLD profile of the PMTAC brush in D_2O was a smooth upward curve from 4.1 $\times$ 10^{-4} nm^{-2} at the Si substrate surface to 6.38 $\times$ 10^{-4} nm^{-2} at the D_2O/ brush

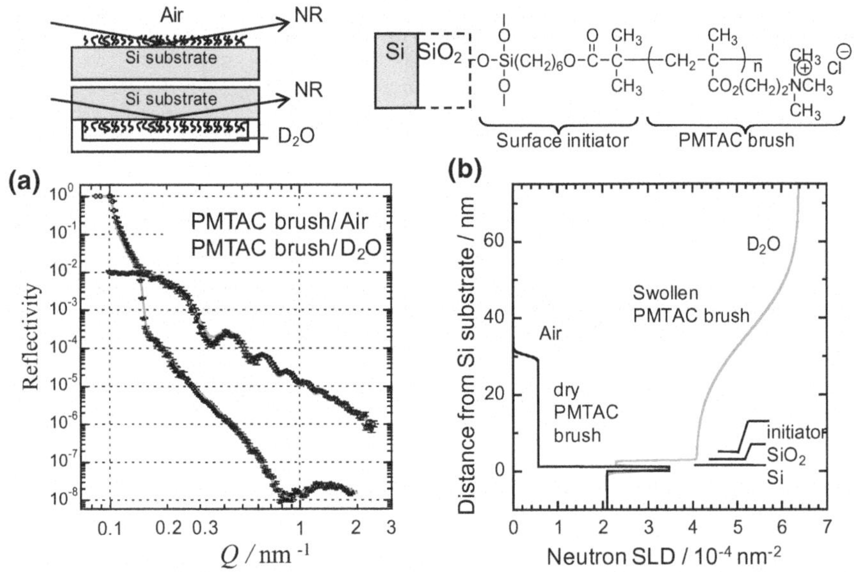

Fig. 8.6 **a** NR curves of the PMTAC brush in air and D_2O, and corresponding **b** neutron SLD profiles along with the distance from the Si disk surface. Reproduced with permission from [6]. Copyright (2013) Nature Publishing Group

interface. The SLD curve of the brush approached SLD_{D2O} at ca. 50 nm from the surface. In general, SLD of the swollen brush at a position (z) apart from the substrate surface is determined by a volume fraction $\phi(z)$ of polymer brush in a solvent. Using SLD of PMTAC and D_2O, SLD(z) can be expressed by

$$\mathrm{SLD}(z) = \mathrm{SLD_{MTAC}}\phi(z) + \mathrm{SLD_{D2O}}(1 - \phi(z))$$

8.4 Grazing Incidence Wide-Angle X-Ray Diffraction

Grazing incidence wide-angle X-ray diffraction (GIXD) is used to study crystalline structure at film surfaces because the X-ray penetration depth (1/e value) is small if the X-ray incident angle is lower than the critical angle. The penetration depth is in the order of nanometer. An evanescent wave is established for a short distance and is exponentially damped. Therefore, Bragg reflections are only yielded from the outermost surface structure. Figure 8.7 shows the schematic geometry of the in-plane GIXD measurement and calculated penetration depth (1/e value) as a function of X-ray incident angle for high-density polyethylene (HDPE). Bragg diffractions from crystallographic planes perpendicular to the film surface are obtained from the surface and bulk regions at the incident angle of X-rays, α_i, of 0.10° and 0.20°, respectively. An advantage of GIXD is that the electric field at the critical angle is amplified locally by a factor of four, making the signal stronger. A disadvantage is the limited in-plane spatial resolution due to the beam footprint.

Figure 8.8 shows the 2-dimensional GIXD pattern of poly(perfluorooctylethyl acrylate) poly(FA-C_8) thin films and the corresponding structure model of the smectic-B phase [7]. The X-ray incident angle α_i was 0.16 degree. A sharp and strong peak was observed at Q_{xy} = ca. 12.5 nm^{-1} in the in-plane GIXD profile. The *d*-spacing calculated from the peak position was ca. 0.50 nm that is close to the intermolecular distance between helical fluoroalkyl (R_f) chains of poly(tetrafluoroethylene) crystals. Because the in-plane GIXD measurements evaluate regularity of the crystallographic plane normal to the film surface, the R_f groups align normal to the film surface. Namely, the rigid rod-like R_f groups produce the hexagonally packed lateral ordered structure. Meanwhile, periodic diffractions are observed in the out-of-plane, indicating the stacked layered lamellar structure of the R_f groups. The highly oriented hexagonally packed R_f groups produce a well-ordered array of CF_3 groups which have extremely low surface free energy leading to the superior liquid repellency [7].

8.5 Conclusions and Perspective

We have presented here the basic principles of reflectivity and examples of XR, NR, and GIXD measurements. XR and NR are powerful to reveal the nanometer-scale

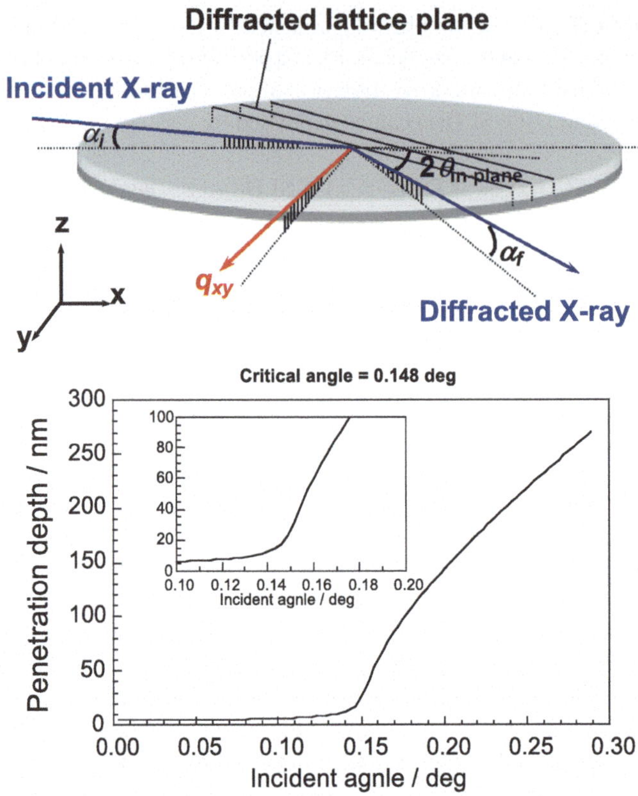

Fig. 8.7 Schematic geometry of the in-plane GIXD measurement, and penetration depth in HDPE films as a function of X-ray incident angle with wavelength of 0.125nm

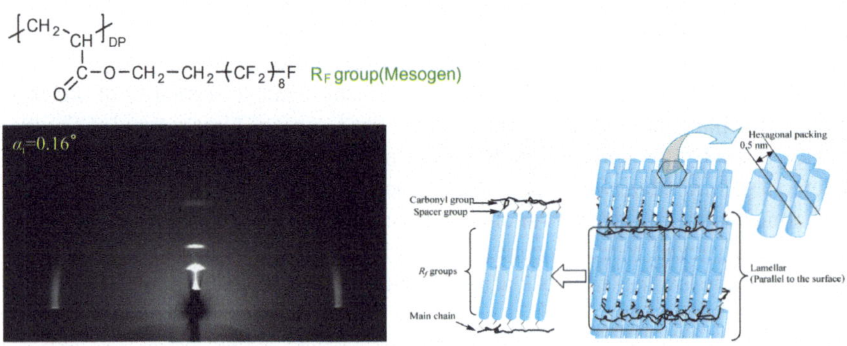

Fig. 8.8 2-Dimensional GIXD pattern of the poly(FA-C_8) thin films, and corresponding structure model of the smectic-B phase. The X-ray incident angle α_i was 0.16°

surface and interface structures including the layered structure, thickness, and interfacial roughness, and these techniques are complement because the contrast between two given elements is different for X-rays and for neutrons. GIXD can be applied for surface sensitive crystalline structure analysis. By the use of 2D detectors, we are able to elucidate the anisotropic ordered structure in the thin films. During the last few decades, these techniques have been quickly expanded to soft-matter scientists and getting popular by the development of user-friendly commercial products. However, we need to understand the principle of the reflectivity and X-ray diffraction to interpret the results properly, while paying careful attention to experimental setup.

Recent advances in XR and GIXD measurements with highly collimated ultrahigh flux third-generation synchrotron radiation X-ray have allowed the rapid and precise thin-film structure analysis. The crystalline structure evolution in the film casting process was studied by in situ GIXD measurements. Also, new-generation neutron facilities with a megawatt-class proton accelerator provide pulsed neutron beam that enables NR measurement by the time-of-flight (TOF) method. This allows us time-resolved, *in situ* measurements of soft materials over a wide Q range. We are able to reveal the structure development process that is triggered by external stimuli such as electric potential, temperature, and chemical substances. Further development in reflectivity and grazing incident X-ray diffraction measurements is anticipated along with the appearance of a new light source.

References

1. Als-Nielsen J, McMorrow D (2001) Elements of modern X-Ray physics. Wiley, New York
2. Russell TP (1990) X-ray and neutron reflectivity for the investigation of polymers. Mater Sci Reports 5:171–271
3. Parratt LG (1954) Surface studies of solids by total reflection of x-rays. Phys Rev 95:359–369
4. Nelson A (2006) Co-refinement of multiple-contrast Neutron/X-Ray reflectivity data using MOTOFIT. J Appl Crystallogr 39:273–276
5. Ogawa H, Masunaga H, Sasaki S, Goto S, Tanaka T, Seike T, Takahashi S, Takeshita K, Nariyama N, Ohashi H, Ohata T, Furukawa Y, Matsushita T, Ishizawa Y, Yagi N, Takata M, Kitamura H, Takahara A, Sakurai K, Tashiro K, Kanaya T, Amemiya Y, Horie K, Takenaka M, Jinnai H, Okuda H, Akiba I, Takahashi I, Yamamoto K, Hikosaka M, Sakurai S, Shinohara Y, Sugihara Y, Okada A (2013) Experimental station for multiscale surface structural analyses of the soft-material Films at SPring-8 Via GISWAX/GIXD/XR integrated system. Polym J 45:109–116
6. Mitamura K, Yamada NL, Sagehashi H, Torikai N, Arita H, Terada M, Kobayashi M, Sato S, Seto H, Gokou S, Furusaka M, Oda T, Hino M, Jinnai H, Takahara A (2013) Novel neutron reflectometer SOFIA at J-PARC/MLF for in-situ soft-interface characterization. Polym J 45:100–108
7. Higaki Y, Ishige R, Takahara A (2014) Handbook of Fluoropolymer science and technology. In: Dennis W, Smith Jr, Iacono ST, Suresh I (eds) fluoropolymer surfaces/interfaces, Chapter 19. John Wiley & Sons Inc., New York, pp. 433–450

Chapter 9
Scanning Electron Microscopy

Xi Jiang, Takeshi Higuchi and Hiroshi Jinnai

9.1 Introduction

Scanning electron microscopy (SEM), an important member of the electron microscopy family, is a versatile instrument widely used in various fields such as nanotechnology, biology, and the life sciences for imaging of micro- and nanostructure morphology and characterizations of chemical composition of various materials. As introduced in the last chapter, Manfred von Ardenne developed electron probe microscopy in 1938 by using a focused electron beam. His work, which is well known as scanning transmission electron microscopy (STEM), could be described as a primitive form of SEM. The first specialized SEM was described in 1942 and the commercial instrument was introduced in 1965 [3, 4]. As a type of electron microscopy, SEM has some principles and features in common with TEM, but there are some differences, too. Both of them use an electron beam to probe the properties and structures in materials. Also, various accessories and special sample stages, e.g., heating, cooling, and in situ mechanical test, could significantly extend the application of SEM as with TEM. In contrast to the thin specimen characterized in TEM by collecting transmitted electrons, SEM uses a focused electron beam to scan the surface of a specimen line by line to acquire information based on interactions between electrons and specimen. It can provide information on surface topography, crystalline structure, chemical composition, and electrical behavior up to several microns deep

X. Jiang
Materials Sciences Division, Lawrence Berkeley National Laboratory, One Cyclotron Road, Berkeley, CA 94720, USA

T. Higuchi
Institute of Multidisciplinary Research for Advanced Materials, Tohoku University, Sendai, Miyagi 980-8577, Japan

H. Jinnai (✉)
Institute of Multidisciplinary Research for Advanced Materials (IMRAM), Tohoku University, 2-1-1, Katahira, Aoba-ku, Sendai, Miyagi 980-8577, Japan
e-mail: hiroshi.jinnai.d4@tohoku.ac.jp

M. Maeda et al. (eds.), *Molecular Soft-Interface Science*,
https://doi.org/10.1007/978-4-431-56877-3_9

from the specimen surface by collecting scattered electrons. The advantages of SEM over TEM include the capacity to image a thick specimen, nondestructive evaluation, and simple operation. The instrument, principles of imaging, and applications will be shortly introduced in the following sections.

9.2 Basic Components of SEM

SEM consists of column, specimen chamber and detectors. SEM column can be divided into the illumination, scanning, and imaging systems, as shown in Fig. 9.1. Components of an SEM column are similar to TEM, but more concise. Primary electrons (PEs) are generated from electron sources (thermionic or field emission) and pass through the anode reaching accelerations up to 30 kV or higher. Once the PEs enters the illumination system, a couple of condenser lenses are used to control the beam current and beam diameter. The next objective lens is used to converge PEs into a fine beam focused onto the surface of the specimen. There are several types of objective lenses, including in-lens, semi in-lens, and out-lens. Selection among these lens types depends on the desired resolution and observed specimen. Scanning coils (deflection coils) are the unique lens located within the objective lens in both x- and y-directions. These coils are usually controlled by sweep voltage and are used to scan the electron beam and change the area to be scanned. Astigmatism, as one of the lens aberrations, is inevitable and difficult to eliminate compared to spherical aberration and chromatic aberration. Therefore, the stigmators are used to compensate this distortion.

In the specimen chamber, the sample stage is installed with a goniometer and other detectors. The goniometer unit allows the sample stage to move in x-, y-, and z-directions and to be tilted in a defined range. The movement in the z-direction is important since the working distance (WD) can be affected. The working distance refers to the distance between the pole piece of the objective lens and the plane where the PEs are focused. The image quality, especially resolution, can be significantly influenced by the WD because it affects the spherical aberration and depth of field. Usually as the WD decreases, the effect of spherical aberration and depth of field will decrease, and the resolution will be improved.

9.3 Various Information Obtainable in SEM

So far, the PEs have been finely focused by lens groups in the column on the way from the electron source to the surface of the specimen. Electron signals generated in SEM are the result of the interactions between PEs and specimen. When the PEs hit the specimen, they are scattered back into vacuum by the interaction with the specimen's atoms as shown in Fig. 9.2(a). The scattered electrons can be defined as one of two types: elastically scattered at a large angle with almost no loss energy and inelastically scattered at a small angle with significant loss energy. Typically, inelastic scattering is the emission of secondary electrons (SE). These inelastic scattered electrons are used for topographical imaging, and characteristic X-rays for chemical composition analysis (Energy dispersive X-ray spectrometry, EDX).

Backscattered electrons (BSE) are electrons that undergo Rutherford processes and leave the sample without notable energy loss and are assumed to be elastic. Some specimens show fluorescence when exposed to an electron beam. The fluorescence produces light photons that can be detected. Therefore, the composition and structure labeled with luminescent molecules can be detected by using cathode luminescence. In the case of a conductive specimen, the PEs that remain inside specimen are detected by the specimen current. The result of the primary beam hitting the specimen is the formation of a teardrop-shaped reaction vessel wherein scattering takes place (Fig. 9.2b). The size of the reaction vessel depends on the accelerating voltage of the PEs and the mean atomic number of the specimen. The reaction vessel will be larger for a higher accelerating voltage, but smaller for a specimen with a higher atomic number. In SEM, the secondary electrons are produced from the surface of the specimen (ca. 10–100 nm), or roughly the top quarter of the reaction vessel; backscattered

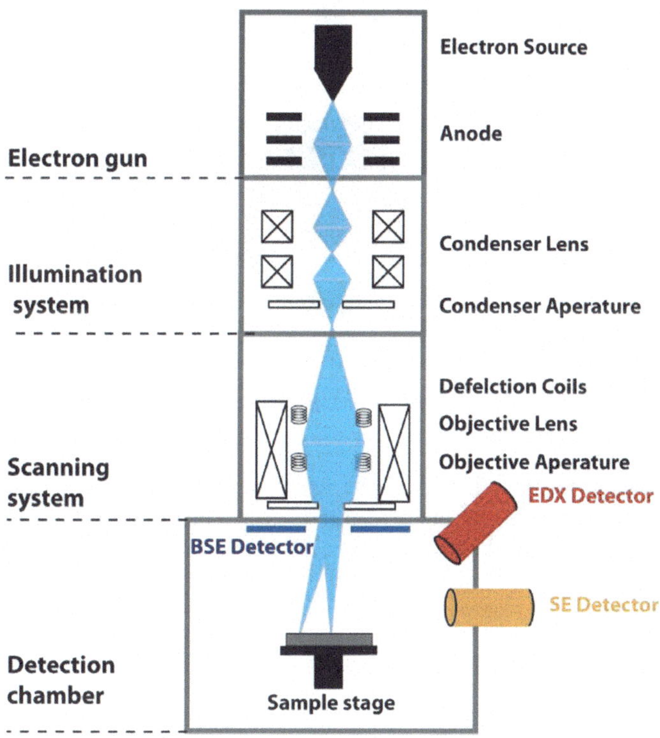

Fig. 9.1 Scheme of scanning electron microscopy

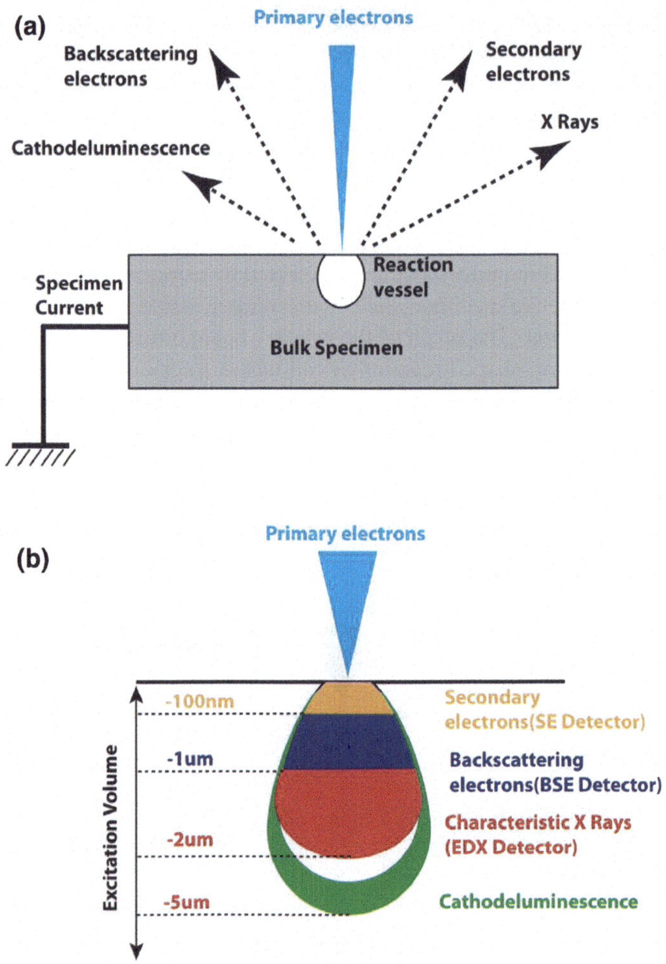

Fig. 9.2 **a** Interactions between incident beam and specimen. **b** Details of scattering in the described reaction vessel

electrons come from the top half of the reaction vessel (up to one micron); and X-rays are generated within the entire reaction vessel. Therefore, the typical thickness that information is detected by SEM can be a few microns.

9.4 A New Type of SEM—Atmospheric SEM

Although SEM has been widely used for investigation of morphology and structures of materials in multiple fields, the limitations of SEM require that the sample should be electrically conductive on the surface, in solid state, and observed in a high vacuum chamber. If the specimen is not electrically conductive, then it needs to be coated with a conductive layer prior to examination in the high vacuum chamber. Hence, environmental scanning electron microscopy (ESEM) was developed to enable uncoated and wet specimens, e.g., live cells, to be observed in their natural states by specially designed detectors and a low vacuum chamber [1]. More recently, the desire to observe in situ physical and chemical phenomena in a liquid or gas has led to atmospheric scanning electron microscopy (ASEM), which was introduced by JEOL [5]. This newly designed instrument combines an inverted SEM at the bottom and invisible light microscopy at the top. The combination of these two techniques records phenomena in a multipurpose dish, which is exposed to atmosphere in real time. Wet materials, which are preserved in natural state, can be observed with ASEM without extensive operator training.

Figure 9.3 shows recent results of a polymeric material using ASEM. Specifically, the system of interest was composed of cationic polymer brushes (poly(2-(methacryloyloxy)ethyltrimethylammonium chloride)) swollen in water and, subsequently, shrank to a dry state. The researchers sought to observe the thickness change of the polymer brushes from their side with ASEM. The polymer brushes were synthesized on lithographically fabricated Au walls. The polymer brush counterions were exchanged from light (Cl-) to heavy atom ions (AuCl4-), which generates electron contrast for cationic polymer brush viewing. Figure 9.3a shows the ASEM images of polymer brushes in the wet state. In this image, the bright region on the left side corresponds to the Au wall. The layer structure was observed, at lower contrast,

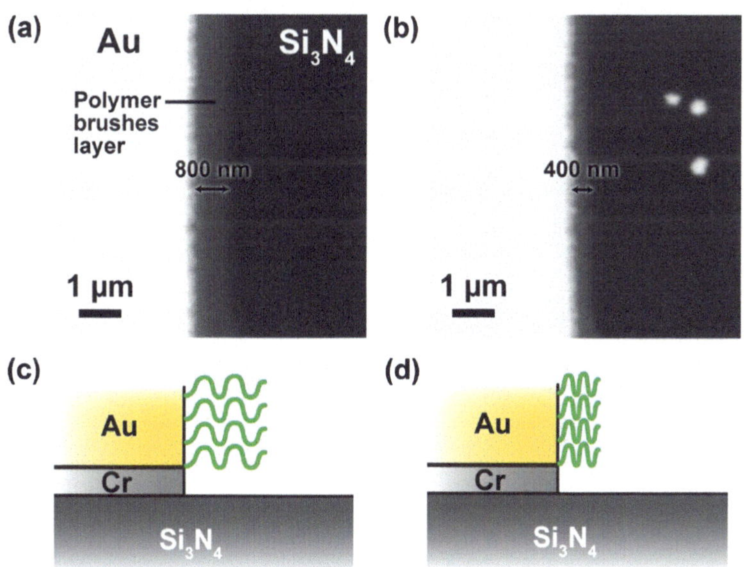

Fig. 9.3 ASEM images of cationic polymer brushes stained with AuCl4 on a Au wall in **a** wet and **b** dry states. Schematic illustrations of cross-sectional view in **c** wet and **d** dry states

to the right of the Au wall, with a thickness of 800 nm. On the other hand, the ASEM image of polymer brushes in the dry state is shown in Fig. 9.3b. The thickness of the polymer brushes was decreased to 400 nm. This result indicates that the swollen polymer brushes shrank by drying (Fig. 9.3c,d). This is the first observation of a shrinkage phenomenon of polymer brushes under atmospheric pressure with electron microscopy [2]. Observation of dynamic phenomena of various chemical and physical reactions in liquid, using ASEM, will provide us with information to clearly support underlying principles.

References

1. Danilatos GD (1988) Foundations of environmental scanning electron microscopy. Adv Electron Electron Phys 71:109–250
2. Higuchi T, Murakami D, Nishiyama H, Suga M, Takahara A, Jinnai H (2014) Nanometer-scale real-space observation and material processing for polymer materials under atmospheric pressure: application of atmospheric scanning electron microscopy. Electrochemistry (in press)
3. McMullan D (1995) Scanning electron microscopy 1928–1965. Scanning 17:175–185
4. Michler GH (2008) Electron microscopy of polymers. Springer Laboratory
5. Suga M, Nishiyama H, Konyuba Y, Iwamatsu S, Watanabe Y, Yoshiura C, Ueda T, Sato C (2011) The atmospheric scanning electron microscope with open sample space observes dynamic phenomena in liquid or gas. Ultramicroscopy 111:1650–1658

Chapter 10
Transmission Electron Microscopy

Xi Jiang, Takeshi Higuchi and Hiroshi Jinnai

10.1 Introduction

Electron microscopy is a versatile scientific technique used in the investigation and characterization of materials science, biology, and life science. The principle of electron microscopy is similar to optical microscopy but uses electrons to illuminate and magnify specimens instead of light. It is well known that the resolution of optical microscopy is limited by the wavelength of visible light. In the early 1930s, the theoretical limitation of visible light microscopy had been reached because of its constant development since the seventeenth century. Electron microscopy was developed so as to overcome the resolution limitation of visible light microscopy and to visualize finer details of materials. Transmission electron microscopy (TEM) was the first form of electron microscopy and was invented by Max Knoll and Ernst Ruska in Germany in 1932 [11]. Ernst Ruska received the Noble Prize for this work in 1986. Scanning transmission electron microscopy (STEM), an important form of TEM, is distinguished from conventional TEM by focusing the electron beam to a spot and scanning across the whole specimen. It was designed by Manfred von Ardenne in Berlin in late 1938 [6, 9]. In the following years, TEM was significantly improved with the advancement of technologies such as the electromagnetic lens, aberration correction, energy filtering, and digital imaging. At the present time, TEM is a reli-

X. Jiang
Materials Sciences Division, Lawrence Berkeley National Laboratory,
One Cyclotron Road, Berkeley, CA 94720, USA

T. Higuchi
Institute of Multidisciplinary Research for Advanced Materials, Tohoku University,
Sendai, Miyagi 980-8577, Japan

H. Jinnai (✉)
Institute of Multidisciplinary Research for Advanced Materials (IMRAM),
Tohoku University, 2-1-1, Katahira, Aoba-ku, Sendai, Miyagi 980-8577, Japan
e-mail: hiroshi.jinnai.d4@tohoku.ac.jp

M. Maeda et al. (eds.), *Molecular Soft-Interface Science*,
https://doi.org/10.1007/978-4-431-56877-3_10

able and powerful tool to investigate morphologies and structures in a wide variety of fields from metallic alloys to inorganics to, soft, and biological materials from micrometer-scale to atomic-scale [2, 4].

Many types of electron microscopy and corresponding functions are derived from modern TEM. High-resolution TEM (HRTEM), which usually equips spherical aberration (Cs) corrector, chromatic aberration (Cc) corrector, and monochromator, refers to directly imaging the atomic structures in a given specimen. It is a powerful tool to investigate properties of materials, such as graphene, nanocatalysts, and quantum dots, at the atomic scale. On the other hand, high contrast pole pieces can be equipped to observe specimens with low contrast for example, biological, medicinal and polymeric materials. Once the TEM, or STEM, combines high-resolution imaging with the analytical capabilities of electron energy loss spectrometers (EELS), energy filters, and energy dispersive X-ray spectroscopy (EDS or EDX) systems, it is the so-called analytical TEM/STEM (AEM). The characterization of structures, chemical composition, and other material properties can be achieved at the atomic-scale with AEM. The TEM specimen holder is another important piece that has the capability to extend functionality. So far, various specimen holders, including cryogenic, heating, cooling, liquid, and straining holders, have been developed to satisfy demands, such as in situ observation of nanocrystal growth [3] and cryogenic observation of biological materials [12]. In the past decade, two additional techniques that have been benefited from digital image acquisition and the advancement of computational power are electron tomography (ET) and electron holography [7, 10, 13]. Originating in the life sciences, ET has grown to usage in materials science. It has been widely used for the 3D morphological and chemical composition characterization of nanostructures. Electron holography provides magnetic and electrostatic information of materials.

10.2 Basic Components of TEM

It is impossible to introduce all of the specific functions and applications of TEM in a few words. However, in general, TEM can be introduced in the following parts in Fig. 10.1: electron gun, illumination system, image creation, and data collection [14]. Figure 10.1 has an energy filter, but it is often optional. The function of TEM is similar to visible light microscopy except that it uses a focused beam of electrons instead of visible light. The basic principle involved in TEM can be described as follows: an electron beam, generated from an electron source under high vacuum, is accelerated toward a thin specimen and then is confined and focused by electromagnetic lenses and apertures into a monochromatic beam.

The electron gun, which is usually located at the top of the instrument, is the first and most basic part of TEM. The two typical types of electron sources are thermionic emission and field emission. Thermionic emission is usually a V-shaped tungsten filament or LaB6 crystals that are wreathed with a Wehnelt cap. In conventional TEM, a positive electrical potential is applied to the anode, and the thermionic emission

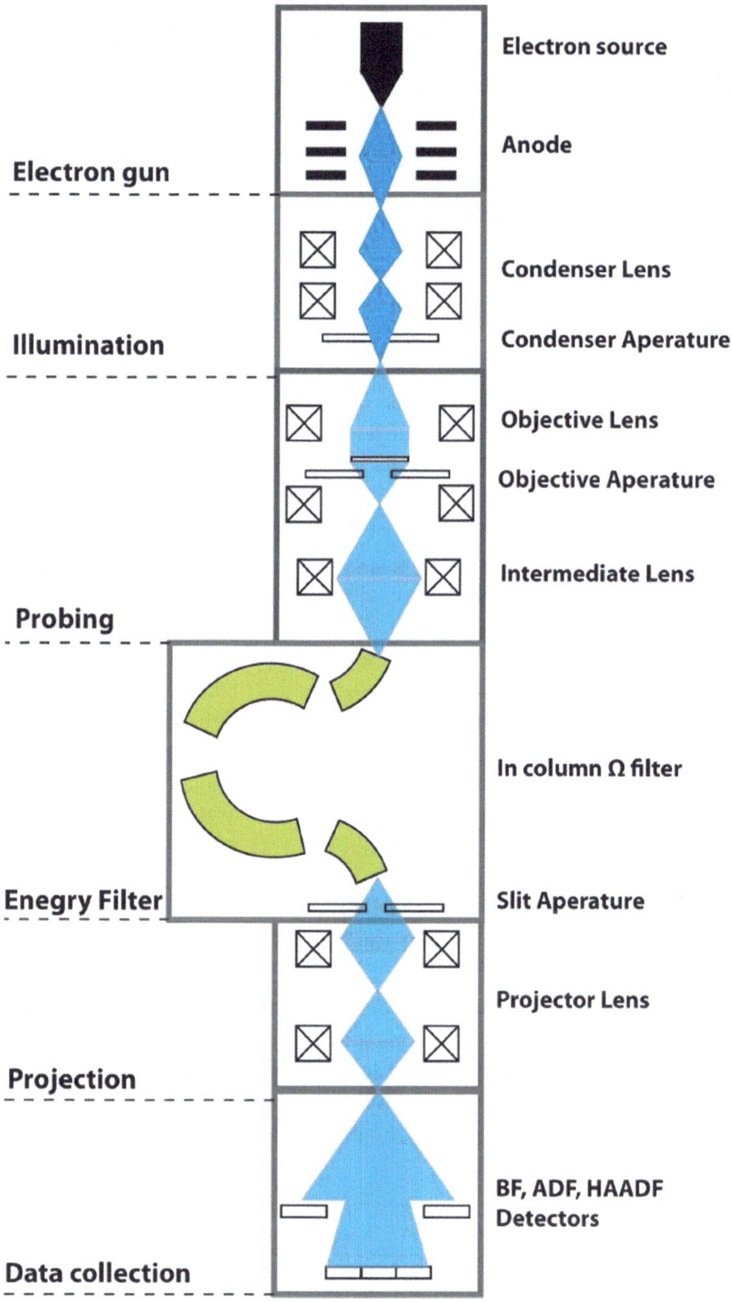

Fig. 10.1 Basic components in TEM and a ray diagram in column. There are more lenses and components in modern TEM

filament (cathode) is heated until an electron beam is produced. The electrons are accelerated to the anode aperture, down the column, and because of the negative potential of Wehnelt cap, all electrons are focused toward the optic axis. A tungsten tip (ca. 100 nm) is used in the field emission type and the cathode is replaced by a pair of anodes. The first anode is positively charged to provide extraction voltage to pull electrons out of the tip, and the second anode accelerates the electron beam to the desired kV. The combined fields of the anodes act like a repelling lens to produce a concentration of electron beam that is similar to a Wehnelt cap. Field emission guns (FEGs), which are widely used in modern TEM, produce higher source brightness and better monochromatic than thermionic guns, but require a very high vacuum and are more expensive.

In the illumination system, the electron beam from the guns crossover is focused on to the specimen by a condenser lens. In modern TEM, usually the illumination part consists of two or more condenser lenses and a condenser aperture to control spot size and beam convergence. The first condenser is used to reduce the crossover and control the minimum spot size obtained in the rest of the illumination part. The second condenser controls convergence of the beam at the specimen and controls the diameter of the illuminated area of the specimen. The condenser aperture controls the fraction of the electron beam that is allowed toward the specimen and controls the intensity of illumination.

To probe and create an image of the specimen's illuminated area, at least three lenses are necessary: an objective lens, an intermediate lens, and a projector lens. The objective lens generates the first intermediate image, which determines the resolution of the final, subsequently magnified image. The intermediate lens magnifies the first intermediate image or diffraction pattern formed in the back focal plane of the objective lens. The magnified images are enlarged by the projector lens and are displayed on the viewing screen or captured by CCD camera.

The energy filter has developed as a component of modern TEM and is highly sensitive with an energy resolution <1 eV. There are two types of energy filters, in-column filter (Omega filter, Zeiss, JEOL) and post-column filter (Gatan imaging filter, Gatan). The in-column, Ω energy filter is placed in the TEM column between the intermediate and projector lenses. It consists of a group of magnetic prisms arranged in an Ω shape that induces the electrons off axis and brings them back to the optic axis before entering the projector lens. Electrons traveling through the spectrometer with particular energy can be selected by the slit aperture. Thus, only electrons of the selected energy range, determined by the slit width, are used to form the image projected onto the detectors.

10.3 Various Modes in TEM Observations

In general, TEM is able to explore various electron beam–specimen interactions including transmitted electrons (in case of TEM mode), elastically scattered electrons (in diffraction mode) and inelastically scattered electrons (in EELS and EFTEM

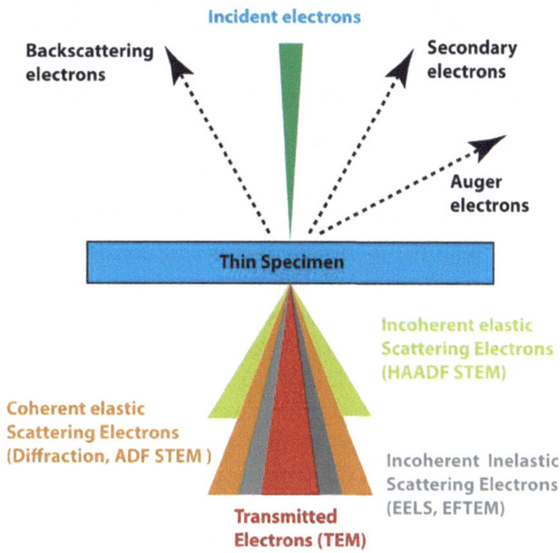

Fig. 10.2 Different kinds of electron scattering and detection modes

modes) as Fig. 10.2 shows. The bright field (BF) detector is placed at the perpendicular site from the aperture in TEM and detects the intensity of the transmitted electrons from a point on the specimen. The annular dark field (ADF) detector is a halo that encompasses the BF detector. The ADF detector collects coherently scattered electrons for image formation, electron diffraction, and STEM, similar to the DF mode in TEM. The high-angle annular dark field (HAADF) detector is also a halo with a hole at the prior position in the column, but the diameter and the hole are much larger than in the ADF and BF detectors. Thus, it detects incoherent elastically scattered electrons, which are scattered to higher angles and leads to a strong atomic number (Z) contrast image.

As we discussed above, TEM provides several types of images and information of a specimen at the atomic-scale. It has been proven as a versatile and powerful analytical tool to explore nanostructures and has shown continuous improvement in past decades. The extension of TEM to ET has proved effective in revealing the nanostructures in various materials including soft material, renewable energy material, and catalytic materials [5]. In the case of ET, projected images of the specimen are recorded at different angles by tilting it with respect to the electron beam in the TEM column. The tilting series are precisely aligned for tilting axis, and then, the 3D images are reconstructed with algorithms such as filtered back projection (FBP) and the simultaneous iterative reconstructive technique (SIRT). An ET observation exercises its great potential for complex structures, which are difficult to determine actual 3D structures from the conventional 2D projected image.

10.4 An Example of Visualization of Soft Interface by TEM

Herein, we introduce an example that highlights ET's ability to reveal the complex polymer interfaces formed by linear triblock terpolymers having polystyrene (PS), polybutadiene (PB), and poly(methyl methacrylate) (PMMA) blocks (PS-*b*-PB-*b*-PMMA). Block copolymers spontaneously form various microphase-separated structures with nanoscale periodicity due to their low compatibilities among constituent blocks. With the increasing number of constituent blocks, resulting morphologies become more complex. For instance, linear triblock terpolymers form intriguing complex structures [1]. Figure 10.3a shows the TEM image of PS-*b*-PB-*b*-PMMA films stained with OsO_4, which reacts with double bonds in PB blocks. The dark regions in the TEM image correspond to the stained PB microdomains. Striped patterns of PB microdomains were observed. The reconstructed 3D image of PB helical microdomains obtained by ET is overlaid on the corresponding region, in which individual PB domains are colored with red and blue. From the 3D structures, the PS-*b*-PB-*b*-PMMA forms a double helical structure of PB blocks surrounding a hexagonally packed PS cylinder in a PMMA matrix and the detailed structural parameters are quantitatively characterized as shown in Fig. 10.3b. Furthermore, the ET observation revealed the handedness of PB helical domains, which could not be determined from conventional 2D projected images.

The HAADF STEM and EFTEM-based ET are advanced techniques, which simultaneously record 3D morphological and compositional information by choosing a slit that corresponds to an energy loss within a particular set of atomic numbers.

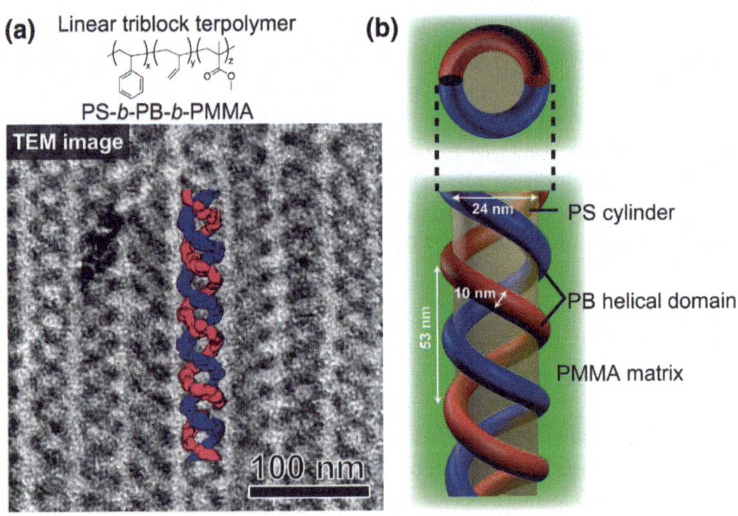

Fig. 10.3 Different kinds of electron scattering and detection modes

These techniques will significantly deepen insights, not only of morphologies, but also atomic composition in a specimen [8]. Further development of ET would achieve an extension to 4D, indicating the time evolution of 3D structures.

References

1. Abetz V, Goldacker T (2000) Formation of superlattices via blending of block copolymers. Macromol Rapid Commun 21:16–34
2. Bethge H (1982) Electron microscopy–beginnings and present. Ultramicroscopy 10:181–186
3. Giorgio S, Joao SS, Nitsche S, Caudanson D, Sitja G, Henry CR (2006) Environmental electron microscopy (etem) for catalysts with a closed e-cell with carbon windows. Ultramicroscopy 106:503–507
4. Haguenau F, Hawkes PW, Hutchison JL, Satiat-Jeunemaitre B, Simon G, William DB (2003) Key events in the history of electron microscopy. Microsc Microanal 9:96–138
5. Jinnai H, Spontak RJ (2009) Transmission electron microtomography in polymer research. Polymer 50:1067–1087
6. McMullan D (1995) Scanning electron microscopy 1928–1965. Scanning 17:175–185
7. Midgley PA, Dunin-Borkowski RE (2009) Electron tomography and holography in materials science. Nat Mater 8:271–280
8. Midgley PA, Weyland M (2003) 3D electron microscopy in the physical sciences: the development of Z-contrast and eftem tomography. Ultramicroscopy 96:413–431
9. Pennycook SJ, Nellist PD (1995) Scanning transmission electron microscopy: imaging and analysis. Springer, New York
10. Rosier DJD, Klug A (1968) Reconstruction of three dimensional structures from electron micrographs. Nature 217:130–134
11. Ruska E (1987) The development of the electron microscope and of electron microscopy. Rev Mod Phys 59:627–638
12. Stahlberg H, Walz T (2008) Molecular electron microscopy: state of the art and current challenges. ACS Chem Biol 3:268–281
13. Tonomura A (1987) Applications of electron holography. Rev Mod Phys 59:639–667
14. Williams DB, Carter CB (1995) Transmission electron microscopy. Springer, New York

Chapter 11
Scanning Probe Microscopy (SPM)

Yoshihiro Kikkawa and Reiko Azumi

11.1 Introduction

Scanning probe microscopy (SPM) detects the local physical quantity (magnitude of the interaction) by using a sharp probe tip that scans sample surfaces. The vertical (Z direction) motion of a piezoelectric device is controlled by a feedback loop to keep the physical quantity between the probe tip and sample surface constant. By adding in-plane motion (XY direction) to the piezo-scanner (or probe tip), namely probe tip scans on a sample surface in two-dimensions, and surface information (e.g., electron density of states, topography, etc) can be mapped as a three-dimensional SPM image at an atomic resolution. The local physical quantity that can be applied for SPM imaging includes tunneling current, attraction and repulsive force, friction, viscoelasticity, and so on.

The history of SPM starts from the invention of scanning tunneling microscopy (STM) in 1982 by Gerd Binnig and Heinrich Rohrer [1], who are the Nobel laureates in Physics in 1986. The STM was initially developed to study the local electrical properties of thin insulating layers. Since the imaging principle of STM is based on the tunneling current between the probe tip and the sample possessing an electrical conductivity, the nonconductive materials such as organic and polymeric materials as well as biological specimens could not be observed. Due to the strong demands for imaging the surface of nonconductive materials, Binnig and his colleagues developed atomic force microscopy (AFM, also known as scanning force microscopy SFM) in 1986 on the basis of the STM principles [2]. AFM provides a three-dimensional surface profile without vacuum condition and pre-sample treatment such as metal coating, which is usually necessary for conventional electron microscopy.

Y. Kikkawa (✉) · R. Azumi
National Institute of Advanced Industrial Science and Technology (AIST), 1-1-1 Higashi, Tsukuba, Ibaraki 305-8565, Japan
e-mail: y.kikkawa@aist.go.jp

M. Maeda et al. (eds.), *Molecular Soft-Interface Science*,
https://doi.org/10.1007/978-4-431-56877-3_11

In this chapter, we introduce the basic principle of SPM (especially for STM and AFM) and application of SPM techniques to the studies of soft interface science and illustrate the availability of SPM for the soft material researches.

11.2 Scanning Tunneling Microscopy (STM)

For the STM studies, a probe tip is made of metals such as tungsten and platinum alloys, and it scans conductive metal surfaces with the help of piezoelectric devices. Once the sharp probe tip approaches very close to the sample surface (a few Å) and a bias voltage is applied, tunneling current passes through the barrier between them by a quantum mechanical process. The direction of electron flow depends on the bias polarity applied to the system. The tunneling current (I) is expressed as follows:

$$I \propto \exp\left(-A\phi^{1/2}s\right)$$

where A is constant (1.025 Å^{-1} $\text{eV}^{-1/2}$), ϕ is average barrier height between the two electrodes (~4 eV), and s is gap width (corresponding to the distance between the probe tip and the sample surface). Roughly, when the gap s varies by 1 Å, the tunneling current I changes by one order of magnitude. Such strong dependence on the gap distance s is the key factor for atomic resolution in the vertical direction.

The STM has two operation modes: constant height and constant current modes. In constant height mode, the vertical position of the probe tip is fixed (feedback signal is OFF), and the tunneling current, which varies due to the topography and local surface electric properties of the sample, is monitored as a function of lateral position. The constant height mode can operate faster because there is no up and down motion of piezo-scanner, but its use is limited to the samples with a relatively smooth surface and small scan area. In the case of the constant current mode, the position of the probe tip is controlled by the piezo-scanner (feedback signal is ON), and the probe tip traces the sample surface at a constant value of tunneling current. The obtained topographic information includes not only physical height but also electron density of states at the surface. The motion of the piezoelectric devices constitutes the image data. This mode can be applied for the irregular surface, but the measurement requires more time due to the feedback loop, whose parameters should be set appropriately not to introduce a periodic noise in the data.

Initially, STM has been applied to the studies of semiconductor, inorganic molecules, and metal surfaces [3–5]. As for the organic materials, self-assembled monolayers (SAM) have been the target for STM analyses [6, 7]. The chemisorbed SAM has been useful for the creation of soft interface with different surface properties and functions (see Chap. 3). The most popular SAM can be fabricated with organic molecules containing thiol group on atomically flat Au(111) surface, which is obtained by vacuum deposition of Au on a freshly cleaved surface of mica heated at ca. 500 °C. In the simple case of alkanethiol ($CH_3(CH_2)_nSH$), STM observa-

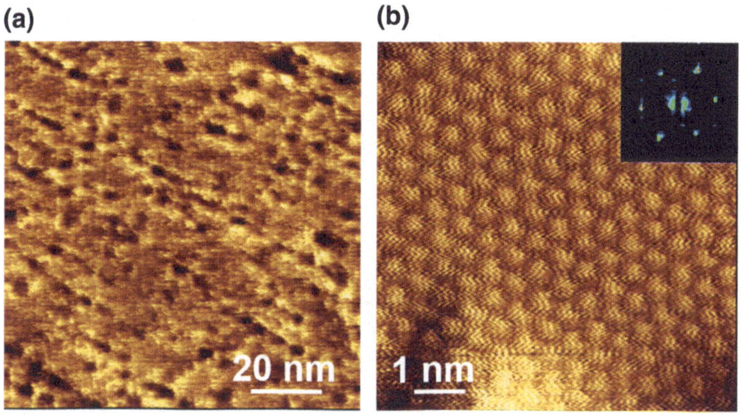

Fig. 11.1 STM images of 1-decanethiol SAM on Au(111). Panels **a** and **b** shows the large scale and magnified **b** STM images, respectively. Inset in **b** shows FFT spectrum of image **b**, indicating the threefold symmetry

tion revealed various crystalline packing structures during its growth on Au(111) [8]. Figure 11.1 shows the example of our STM image of 1-decanethiol monolayer with the $\left(\sqrt{3} \times \sqrt{3}\right)$ R30° structure, which was formed from an ethanol solution on Au(111). The SAM is composed of all-trans alkyl chains tilted ca. 30° against the substrate normal axis as well as etch pits.

In physisorbed SAM, molecules are flatly oriented on a substrate, for which freshly cleaved highly oriented pyrolytic graphite (HOPG) is often used. The lattice constant of HOPG ($a = b = 0.246$ nm, $\gamma = 60°$), which always exists during the imaging, can be applied for the correction of the lateral dimensions in the STM images of the physisorbed organic monolayer. Submolecularly resolved STM imaging of spontaneously formed monolayer has been performed at the solid/liquid interface to study the correlation between the molecular structures and the resultant 2D molecular orientations [9, 10]. Understanding of the supramolecular interaction on a surface is essential for the fabrication of highly ordered two-dimensional patterns, which have potential application to nanoelectronics, molecular devices, and lithographic techniques. A notable feature of the solid/liquid interface system is the availability of post-reaction, such as metallation, thermal treatment, blending, and UV irradiation [11]. Figure 11.2 shows the STM images of thermally responsive isobutenyl compounds with different lengths of alkyl chains at the HOPG/1-phenyloctane interface [12, 13]. Due to the different tunneling efficiency, the naphthalene and alkyl chain units were visualized as bright dots and dark lines, respectively. The isobutenyl compound possessing C18 alkyl chains displayed wavy structure, whereas that with C19 alkyl chains formed C_3 symmetric tripod structure. Thus, the 2D structural features are governed by the odd-even effect of alkyl chains. After the thermal treatment of the compounds, tandem Claisen rearrangement (TCR) occurred, resulting in the transformation from ether to hydroxyl groups accompanied by a new C–C bond

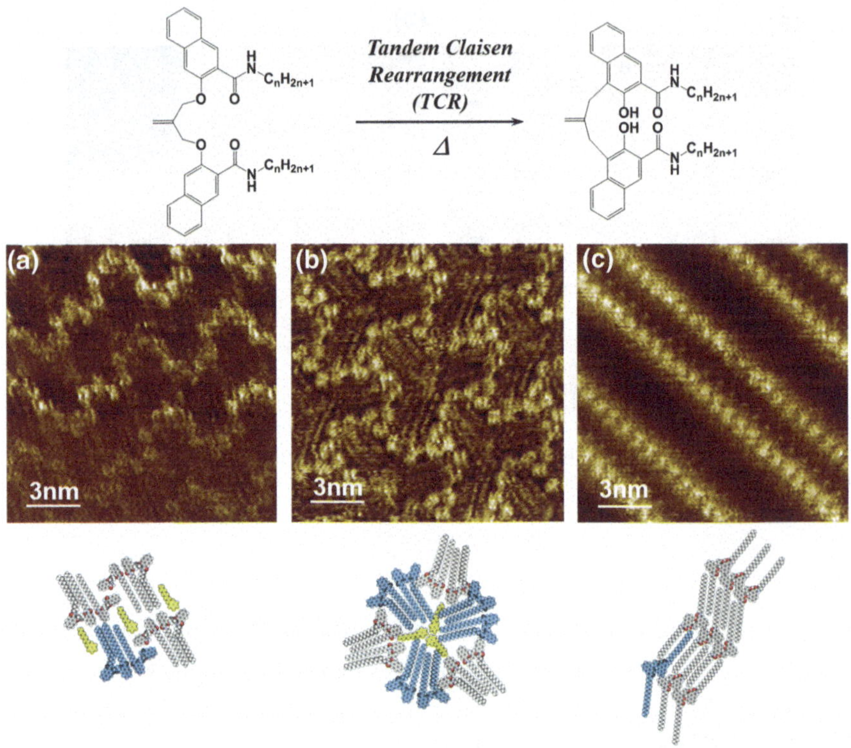

Fig. 11.2 Chemical structures of isobutenyl compounds before and after the TCR. Panels **a** and **b** show the STM images of the isobutenyl compounds possessing C18 and C19 alkyl chains, whereas panel **c** shows the thermally treated isobutenyl compound with C18 alkyl chains. The STM observation was performed at the HOPG/1-phenyloctane interface. The molecular models for each STM images are shown under the STM images

formation. Then, the STM observation revealed that linear structure was formed, regardless of the alkyl chain length, namely, odd–even effect was canceled due to the TCR.

11.3 Atomic Force Microscopy (AFM)

AFM can be used for both conducting and nonconducting surfaces. A probe tip is generally made of Si_3N_4 or Si, and is mounted at the end of the flexible cantilever. The probe tip radius typically ranges from 5 to 20 nm, and the lateral resolution strongly depends on it. The deflection of the cantilever is detected as the position of the reflected laser beam at the photodetector (Fig. 11.3). The signal from the photodetector passes through a feedback loop to control the z position of the scanner,

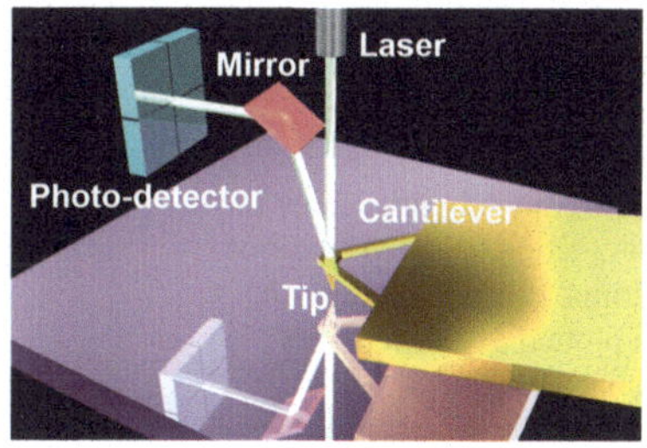

Fig. 11.3 Schematic drawing of AFM setup

and the probe tip-sample force is maintained constant. Addition of lateral scanning motion to the cantilever tip allows to monitor the surface topographic information.

Among various AFM imaging modes, contact, and dynamic force (tapping) modes are generally utilized for the visualization of soft material's surface. Each mode has its specific advantages and disadvantages, and therefore the selection of the appropriate imaging mode is important for the nondestructive and high-resolution imaging of samples. The interaction force between the probe tip and sample surface can be expressed by proximity of the Lennard–Jones potential, which represents the interaction energy between two adjacent atoms or molecules. The potential energy $E(r)$ is given by the following equation:

$$E(r) = -4\varepsilon\left[\left(\frac{\sigma}{r}\right)^{6} - \left(\frac{\sigma}{r}\right)^{12}\right]$$

where ε is the energy minimum, r is the distance between the two objects (the probe tip and the sample surface in the present case), σ is the distance at which the $E(r)$ is zero. As shown in Fig. 11.4, the interaction is an attractive force in the region of $r > r_0$ due to van der Waals interaction, whereas in $r < r_0$, repulsive force works between the probe tip and the sample surface. In the contact mode, the probe tip contacts directly to the sample surface in the repulsive force region, which is maintained during the scan. By using the contact mode in liquid to eliminate the capillary force existing at an ambient condition, there are some reports on high-resolution imaging of biological samples such as ATP synthase, connexons, and so on [14]. However, the lateral force due to the scanning motion of the probe tip sometimes gives damages to the surface of soft materials, and this point should be taken into account for the selection of imaging mode as well as the setting of imaging parameters including feedback gain, scanning rate, and contact force.

The most frequently used AFM imaging mode is the tapping mode (intermittent contact mode), in which the cantilever is oscillated near its own resonance frequency by a small piezoelectric device mounted on the AFM tip holder. As a literal meaning of "tapping," the probe tip moves up and down in the order of kHz frequency, resulting in the elimination of destructive lateral and capillary forces on a sample surface. In this mode, the typical cantilever oscillation amplitude is tens of nanometers, which is kept at a constant value by a feedback loop during the scan on a surface. The

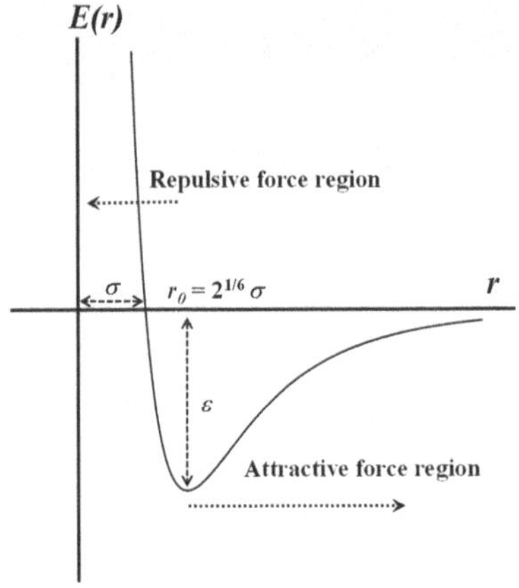

Fig. 11.4 Schematic diagram of Lennard–Jones potential, which is the mathematical model approximately describing the interaction between a pair of atoms. Potential energy (E) is plotted as a function of separation (r)

interaction force between the probe tip and the sample surface can be controlled by changing the driving amplitude (A_0) and set-point amplitude (A_{st}). Careful parameter setting for tapping mode operation allows eliminating the sample damage, which is often the problem for contact mode.

The tapping mode is suitable for imaging the surface morphology of soft materials including organic and polymeric materials as well as biological samples such as proteins and DNA [15]. Among a lot of publications containing tapping mode AFM images, two of our studies are illustrated. The first example is the enzymatic degradation of biodegradable poly[(*R*)-3-hydroxybutyrate] (PHB) copolymer [16]. Thin film of Poly[(*R*)-3-hydroxybutyrate-co-10 mol%-6-hydroxyhexanoate] (P(3HB-*co*-6HH)) was immersed in a buffer solution, and the enzymatic degradation reaction was initiated by adding the PHB depolymerase from *Ralstonia pickettii* T1. As shown in Fig. 11.5, the time-dependent enzymatic degradation process was monitored in situ by using the tapping mode AFM in the buffer solution containing PHB depolymerase. The size of lamellar crystals got smaller due to the enzymatic hydrolysis, and finger-like morphologies were gradually emerged along the long axis of the crystal. This result suggests that the lamellar crystals of P(3HB-*co*-6HH) are composed of tightly and loosely chain-packing regions along the crystallographic *a*-axis, and that the latter disordered regions are preferentially eroded by the PHB depolymerase.

The second example is the tapping mode AFM observation of rings and coils formed via supramolecular self-assembly, as shown in Fig. 11.6 [17]. The building block was composed of barbituric acid head group, oligo(*p*-phenylenevinylene) π-conjugated unit and tridodecyloxybenzyl tail. A methylcyclohexane solution of the building block was drop-cast on HOPG. Nanodonuts with ca. 40 nm in diameter was formed at the relatively low concentration (20 μM), whereas nanorods and open-ended nanofibers were observed in addition to the ring-shaped objects at the higher concentration (100 μM). These structures were proposed to consist of the stacked

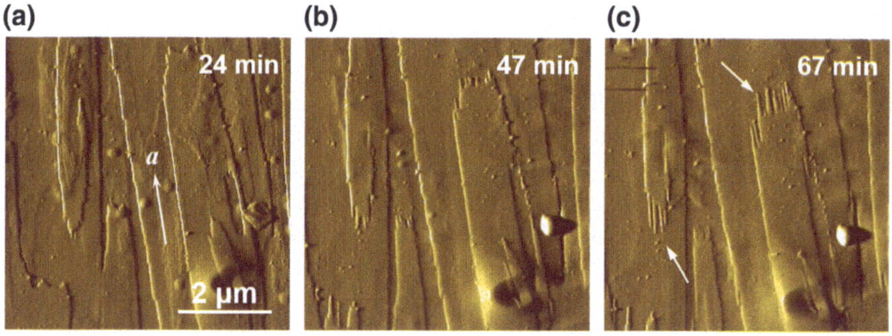

Fig. 11.5 Time-dependent AFM images of the P(3HB-*co*-6HH) thin film during enzymatic degradation in the buffer solution containing PHB depolymerase from *R. pickettii* T1. The images were taken during enzymatic degradation for 24 min **a**, 47 min **b,** and 67 min **c**, respectively. The arrow in the panel **a** indicated the direction of the crystallographic *a*-axis of the lamellar crystals. In panel **c**, finger-like morphology is clearly visible at the tip of crystals indicated by arrows

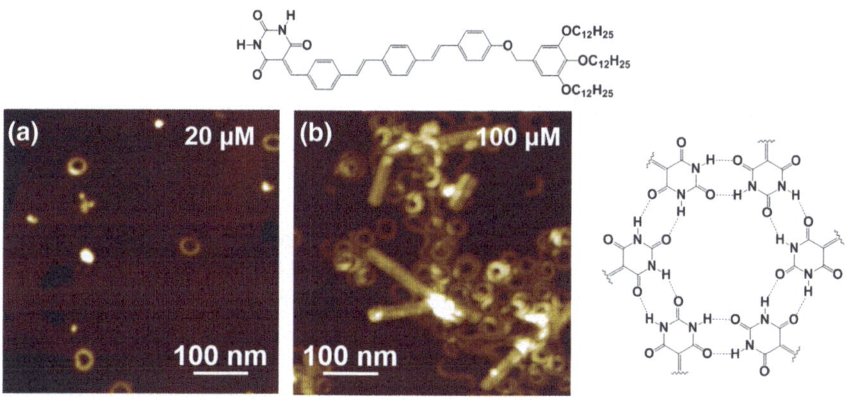

Fig. 11.6 Chemical structure of a molecule with a barbituric acid head group and AFM images of the self-assembled nanodonuts and nanorods. In panel **a**, nanodonuts were found at the concentration of 20 μM, whereas in panel **b**, there are open-ended fibers and nanorods in addition to the nanodonuts. Thus, the supramolecular structures could be controlled by the solution concentration

hexameric rosettes formed by the hydrogen bonding of barbituric acid head groups. Thus, the morphological features could be tuned dependent on the concentration of the building blocks.

11.4 Other Scanning Probe Techniques

In addition to the topographic imaging, SPM has the capability of measuring the interaction force between the probe tip and the sample surface. The interaction forces include not only repulsive and attractive forces, but also friction, viscoelasticity, etc. These SPM techniques are called as force spectroscopy, friction force microscopy (FFM), viscoelastic mode AFM (VE-AFM), respectively.

In the force spectroscopy, the cantilever position in the XY plane is fixed, and the deflection of the cantilever (Z direction) during the up and down motion of the sample on a piezo-scanner is monitored. The interaction force (F) can be calculated according to the Hooke's law ($F = kx$ where k is the spring constant of the cantilever; x is the cantilever deflection). F is plotted as a function of the Z-piezo position (displacement of the scanner), affording a force-displacement curve (force curve). The cantilever motion and the corresponding force curve are schematically depicted in Fig. 11.7. The sample approaches to the cantilever tip (1); the tip jumps into contact to the sample surface when the attractive force (capillary force in air) overcomes the spring constant of the cantilever (2); further movement of the sample causes a cantilever deflection, and repulsive force increases (3); during the retraction of the sample, the tip maintains contact with the sample surface as long as the interaction force works (4); finally, the tip is detached from the surface. Thus, the attractive and repulsive forces between the cantilever tip and sample surface can be estimated from the force curve. Functionalization of the cantilever tip with specific chemicals or enzymes allows to measure the interaction forces between complimentary receptor-ligand systems such as deoxyribonucleic acid (DNA) strands [18], antibody-antigens [19], supramolecular host-guest complexes [20], enzyme-polymer [21], etc. Mapping of chemical species on a surface by using a functionalized cantilever tip is specially called as chemical force microscopy (CFM).

FFM reflects the lateral deflection of the cantilever tip. The FFM signals are derived from the surface characteristics due to the difference in frictional coefficient

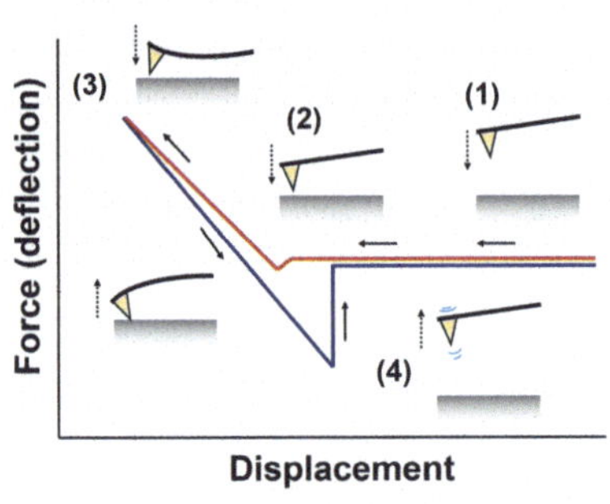

Fig. 11.7 Schematic representation of force curve. The distance between the sample surface and the cantilever tip becomes closer (1); the cantilever tip touches on the sample surface (2); the cantilever is vent and then the tip is pulled up (3); finally, the tip is detached from the sample surface after overcoming the interaction force between the tip and the sample surface

as well as the onset of height changes. FFM observation revealed the direction of chain-folding in a polymer single crystal [22, 23]. Solution-grown single crystal of polyethylene is lozenge-shaped with the {110} growth faces, where the chain direction is parallel to the growth plane in each sector. Depending on the scanning direction, two of the four sectors showed different contrast because of the difference in chain folding direction [24].

The surface molecular mobility of polymeric materials has been studied by using FFM and VE-AFM [25, 26], in which modulation of sample scanner leads to the deflection of cantilever and sample deformation, and their extent is dependent on the viscoelastic property of the sample. Drastic alteration of friction force and viscoelasticity of polymeric surface against temperature has a correlation with the glass transition temperature. By using the FFM and VE-AFM, it was found that the glass transition temperature of polymeric materials on a surface is much lower than that in the bulk measured by conventional differential scanning calorimetry (DSC).

A brief overview of SPM techniques was provided in this chapter. SPM can "visualize" the topography and "measure" the interaction force. The SPM is now widely accepted as the versatile technique for the study of soft-interface science, some of which were exampled above. For more detailed principles, applications and family of SPM (Kelvin force microscopy, Magnetic force microscopy, Near-field scanning optical microscopy, etc) omitted in this chapter, see the consummated reviews [15, 27, 28].

References

1. Binnig G, Rohrer H, Gerber C, Weibel E (1982) Surface studies by scanning tunneling microscopy. Phys Rev Lett 49:57–61
2. Binnig G, Quate CF, Gerber C (1986) Atomic force microscopy. Phys Rev Lett 56:930–933
3. Coratger R, Sivel V, Ajustron F, Beauvillain J (1994) Scanning tunneling microscopy and related techniques for surface analysis. Micron 25:371–385
4. Kubby JA, Boland JJ (1996) Scanning tunneling microscopy of semiconductor surfaces. Surf Sci Rep 26:61–204
5. Hwang RQ, Bartelt MC (1997) Scanning tunneling microscopy studies of metal on metal epitaxy. Chem Rev 97:1063–1082
6. Poirier GE (1997) Characterization of organosulfur molecular monolayers on Au(111) using scanning tunneling microscopy. Chem Rev 97:1117–1128
7. Smith RK, Lewis PA, Weiss PS (2004) Patterning self-assembled monolayers. Prog Surf Sci 75:1–68
8. Noh J, Hara M (2002) Final phase of alkanethiol self-assembled monolayers on Au(111). Langmuir 18:1953–1956
9. Cyr DM, Venkataraman B, Flynn GW (1996) STM investigations of organic molecules physisorbed at the liquid–solid interface. Chem Mater 8:1600–1615
10. De Feyter S, De Schryver FC (2005) Self-assembly at the liquid/solid interface: STM reveals. J Phys Chem B 109:4290–4302
11. Kikkawa Y (2013) Formation of 2D structures and their transformation by external stimuli: a scanning tunneling microscopic study. Polym J 45:255–260

12. Omori K, Kikkawa Y, Kanesato M, Hiratani K (2010) Fabrication and transformation of novel two-dimensional tripod structures: structural modulation by alkyl chain length and tandem claisen rearrangement. Chem Commun 10:8008–8010
13. Kikkawa Y, Omori K, Takahashi M, Kanesato M, Hiratani K (2012) Effects of alkyl chain length, solvent and tandem claisen rearrangement on two-dimensional structures of noncyclic isobutenyl compounds: scanning tunneling microscopic study. Org Biomol Chem 10:8087–8094
14. Müller DJ, Anderson K (2002) Biomolecular imaging using atomic force microscopy. Trends Biotech 20:S45–S49
15. Morris VJ, Kirby AR, Gunning AP (1999) Atomic force microscopy for biologists. Imperial College Press, London
16. Kikkawa Y, Hirota T, Numata K, Tsuge T, Abe H, Iwata T, Doi Y (2004) In situ atomic force microscopy observation of enzymatic degradation in poly (hydroxyalkanoic acid) thin films: normal and constrained conditions. Macromol Biosci 4:276–285
17. Yagai S, Kubota S, Saito H, Karatsu T, Kitamura A, Kanesato M, Kikkawa Y (2009) Reversible transformation between rings and coils in a dynamic hydrogen-bonded self-assembly. J Am Chem Soc 131:5408–5410
18. Lee GU, Chrisey LA, Colton RJ (1994) Direct measurement of the forces between complementary strands of DNA. Science 266:771–773
19. Hinterdorfer P, Baumgartner W, Gruber HJ, Schilcher K, Schindler H (1996) Detection and localization of individual antibody-antigen recognition events by atomic force microscopy. Proc Natl Acad Sci USA 93:3477–3481
20. Zou S, Schönherr H, Vancso GJ (2006) Atomic force microscopy-based single-molecule force spectroscopy of synthetic supramolecular dimers and polymers. In: Samorì P (ed) Scanning probe microscopies beyond imaging, manipulation of molecules and nanostructures. Wiley-VCH, Weinheim, pp 315–353
21. Kikkawa Y, Fukuda M, Kimura T, Kashiwada A, Matsuda K, Kanesato M, Wada M, Imanaka T, Tanaka T (2014) Atomic force microscopic study of chitinase binding onto chitin and cellulose surfaces. Biomacromol 15:1074–1077
22. Kajiyama T, Ohki I, Takahara A (1995) Surface morphology and frictional property of polyethylene single crystals studied by scanning force microscopy. Macromolecules 28:4768–4770
23. Iwata T, Doi Y (2000) Morphology and crystal structure of solution-grown single crystals of poly[(*R*)-3-hydroxyvalerate]. Macromolecules 33:5559–5565
24. Ryousho Y, Sasaki S, Nagamura T, Takahara A, Kajiyama T (2004) Anisotropic-to-isotropic change in lateral force at the surface of single-crystal lamellae of high-density polyethylene during low-temperature annealing. Macromolecules 37:5115–5117
25. Tanaka K, Takahara A, Kajiyama T (2000) Rheological analysis of surface relaxation process of monodisperse polystyrene films. Macromolecules 33:7588–7593
26. Satomi N, Takahara A, Kajiyama T (1999) Determination of surface glass transition temperature of monodisperse polystyrene based on temperature-dependent scanning viscoelasticity microscopy. Macromolecules 32:4474–4476
27. Birgi KS (2003) Scanning probe microscopes: applications in science and technology. CRC Press, Florida
28. Samorì P (ed) (2006) Scanning probe microscopies beyond imaging, manipulation of molecules and nanostructures. Wiley-VCH, Weinheim

Part IV
Application of Soft Interface

Chapter 12
High-Performance Interface

Motoyasu Kobayashi and Atsushi Takahara

12.1 Polymer Brush

Novel and significant applications of soft materials are produced not only by high-performance bulk properties, but also by sophisticated surfaces and interfaces including well-designed molecular structures and nanometer-order morphologies. Adhesion, friction, wettability, and antifouling are all related to the surface and interfaces of the materials. Therefore, various design and modification methods for soft interfaces have attracted much attention for practical applications.

One of the most useful methods for surface modification is grafting a polymer that differs from the base materials. Surface-tethered polymers with sufficient graft density are called polymer brushes [1, 2]. By definition, a polymer brush can be described as polymer chains tethered to a surface or interface with a sufficiently high graft density such that the chains are forced to stretch away from the tethering site [3]. Owing to the covalent immobilization of the brush, the surface properties can be maintained permanently because the brush chains cannot be removed from the substrates by solvent washing or friction. This is a large difference from conventionally coated films, such as cast films, which are easily removed from the surfaces in a solvent. Thus, surface grafting has emerged as a simple, useful, and versatile approach to improve surface properties of polymers for a wide variety of applications.

M. Kobayashi
School of Advanced Engineering, Kogakuin University, 2665-1 Nakano-cho Hachioji, Tokyo 192-0015, Japan
e-mail: motokoba@cc.kogakuin.ac.jp

A. Takahara (✉)
Institute for Materials Chemistry and Engineering, Kyushu University, 744 Motooka, Nishi-ku, Fukuoka 819-0395, Japan
e-mail: takahara@cstf.kyushu-u.ac.jp

Japan Science and Technology Agency, ERATO Takahara Soft Interfaces Project, CE80, 744 Motooka, Nishi-ku, Fukuoka 819-0395, Japan

M. Maeda et al. (eds.), *Molecular Soft-Interface Science*,
https://doi.org/10.1007/978-4-431-56877-3_12

Recently, various types of polymer brushes have been fabricated on solid surfaces by "grafting from" methods combined with a controlled polymerization technique. Surface-initiated controlled radical polymerization offers many parameters related to the interface properties, such as chemical structure, thickness and chain length (molecular weight), grafting density, and morphology. Therefore, the surface or interface properties can be precisely controlled at the molecular level by the polymer brush design. This chapter describes the high-performance interfaces achieved by specifically designed polymer brushes on solid surfaces, focusing on antifouling, adhesion, and low friction features [4].

12.2 Wettability and Antifouling

Antifouling and self-cleaning can be found in living systems for sustaining diverse functions. These functions have attracted the attention of surface-related science researchers due to their importance in medical, industrial, and environmental applications. For example, antifouling and self-cleaning surfaces are desired in the fields of biomedical implants, water purification, and marine coatings.

The self-cleaning surfaces have often been prepared by capitalizing on low surface energy characteristics, accomplished via surface coating or grafting of nonpolar hydrophobic polymers such as silicone-based [5] and fluorinated [6, 7] polymers. These superhydrophobic surfaces minimize the intermolecular forces between foulants and the surface such that the foulants are removed by low hydrodynamic shear forces or by simple mechanical cleaning, resulting in desirable "fouling-release" properties [8–10].

The superhydrophilic surfaces have also been utilized in constructing oleophobic and antifouling surfaces in the wetted state by water. Various hydrophilic polymer brushes have been prepared for antifouling surfaces, typically using poly(ethyleneglycol)-based [11–13], zwitterionic [14–17], glycomimetic [18, 19], and peptidomimetic [20, 21] polymers. These hydrophilic surfaces may form a hydration layer on the surface through hydrogen or ionic bonding and effectively prevent the surface nonspecific adsorption of foulants and oil pollutants, thus exhibiting a superior "fouling-resistant" property [22].

Takahara et al. investigated the antifouling properties of various polymer brushes having a wide range of surface free energies [23, 24], such as poly(2-perfluorooctylethyl acrylate) (PFA-C_8) [25, 26], poly[2-(methacryloyloxyethyl)trimethylammonium chloride] (PMTAC) [27], poly(3-sulfopropyl methacrylate potassium salt (PSPMK), and poly(2-methacryloyloxyethyl phosphorylcholine) (PMPC) [28]. Figure 12.1 shows the contact angles of water and hexadecane in air, and air bubbles and hexadecane in water using conventional static contact angle measurements. The polyelectrolyte brushes on silicon substrates showed significantly low water contact angles below 5°. In particular, the surface free energy γ_{SV} of the PSPMK and PMPC brush surfaces were estimated to be 72.9 mJ m^{-2} by the Owens' protocol [29], which

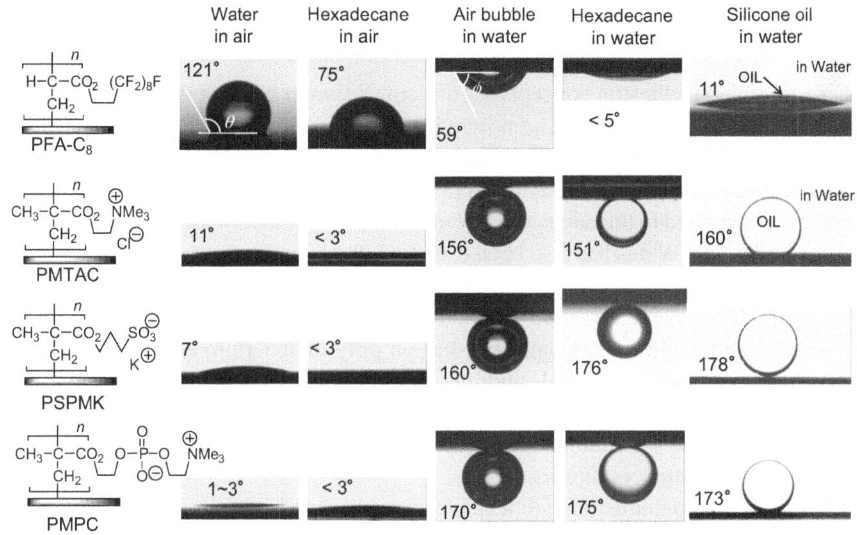

Fig. 12.1 Contact angles of water and hexadecane in air and air bubbles, hexadecane, and silicone oil in water on substrates with PFA-C_8, PMTAC, PSPMK, and PMPC brushes

is quite similar to that of water. Polyelectrolyte brush surfaces repelled both the air bubble and hexadecane in water. The zwitterionic PMPC brush showed particularly excellent antifouling properties. On the other hand, the PFA-C_8 brush showed very low surface free energy but hexadecane was spread at the water interface. This behavior indicated that the hydrophobic surface had high affinity for hydrocarbons.

Polyelectrolyte brushes exhibited antifouling behavior in water against silicone oil [density = 0.965 g/mL at 293 K, interfacial free energy (γ_{OW}) at silicone oil /water = 7.2 mN/m]. In air, the silicone oil droplet is made wet and spread on the surface of the polyelectrolyte brushes. However, once the brush substrate was dipped into water, the silicone oil quickly formed a sphere and eventually detached from the brush surface. The stability of the PMPC brush was also confirmed on a polyolefin sheet [30]. In contrast, the silicone oil still remained attached on the PFA-C_8 brush both in air and water.

The removal of oil drops from solid surfaces immersed in an aqueous medium is of interest in many applications such as self-cleaning, antifogging, and antifouling systems. Oleophobic and antifouling surfaces have been widely fabricated with conventional superhydrophobic materials, such as fluoropolymers, with micro/nanohierarchical structures such as a lotus leaf. In contrast, self-cleaning water/solid interfaces achieved by superhydrophilic polyelectrolyte brush surfaces are proposed here.

12.3 Adhesion

There are several adhesion concepts for utilizing polymer brush surfaces. Ikeda et al. reported in 1994 that hydrophilic polymer brushes underwent substantial adhesion to another surface when they were brought into contact in the presence of water under pressure, and subsequently dried [31]. In this case, interdigitation of opposing brushes was applied to the adhesion. The electrostatic interaction between oppositely charged polyelectrolytes has also been used for adhesion purposes. When oppositely charged polymers contact each other, they form a polyelectrolyte complex [32] due to the strong electrostatic attractive interaction. A pH-sensitive adhesion of poly(*N*,*N*-dimethylaminoethyl methacrylate) brushes on poly(methacrylic acid) gel has been reported by La Spina et al. [33]. Sudre et al. demonstrated controlled adhesion of a poly(acrylic acid) brush on a hydrogel in aqueous solution under various pH conditions [34]. These results indicate that the adhesion force or strength is controllable by the environmental conditions.

Takahara et al. demonstrated repeatable nanoadhesion based on electrostatic interactions between oppositely charged polyelectrolyte brushes combined in deionized water or an aqueous salt solution (Fig. 12.2) [35]. Positively charged PMTAC and negatively charged PSPMK brushes approximately 100 nm thick were prepared by surface-initiated atom transfer radical polymerization on a silicon wafer. Two substrates with oppositely charged polyelectrolyte brushes bonded strongly to each other using a small amount of deionized water to yield a lap shear adhesion strength of 1.52 MPa. Interestingly, the brushes also smoothly debonded in aqueous NaCl solution and each brush layer remained on the substrate after separation because the chain end of the polymer brush was covalently bonded to the solid surface. The positively charged PMTAC brush and the negatively charged PSPMK brush adhered again after washing with deionized water to remove salt.

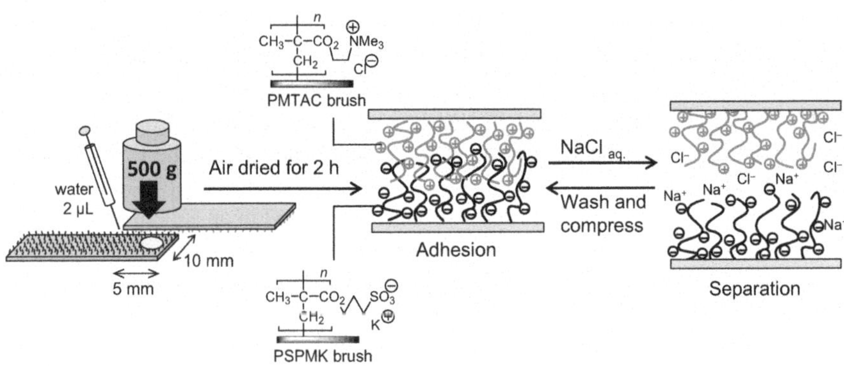

Fig. 12.2 Schematic view of the repeatable adhesion and separation by using oppositely charged polyelectrolyte brushes

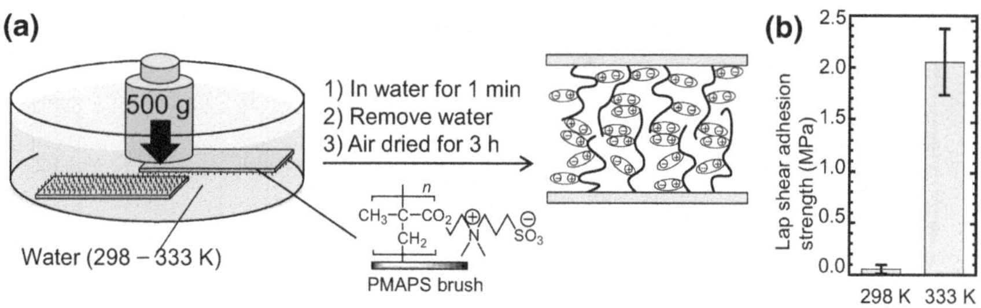

Fig. 12.3 **a** Adhesion procedure of the PMAPS brushes and **b** lap shear adhesion strength of PMAPS brushes in cold water at 298 K and hot water at 333 K

The dipole–dipole interaction between zwitterions is also a good candidate for adhesion. Neoh and coworkers prepared surface-grafted zwitterionic poly[3-(*N*-2-methacryloyloxyethyl-*N*,*N*-dimethyl)ammonium propane sulfonate] (PMAPS) on ozone- or argon-plasma-pretreated polymer films by photolytically or thermally induced graft copolymerization [36–44]. Two polyelectrolyte-grafted films were joined by direct contact in the presence of water and subsequent drying due to the attractive dipole–dipole interaction between the sulfobetaine units. PMAPS has unique physical and chemical properties [45, 46] as well as biomimetic characteristics such as reducing protein adsorption [47, 48] and blood biocompatibility [49, 50].

Takahara et al. also demonstrated that two substrates with PMAPS brushes with 100 nm thickness bonded strongly to each other following compression in deionized hot water and successive air drying to yield a lap shear adhesion strength of 2.05 ± 0.43 MPa, as shown in Fig. 12.3b [51]. However, PMAPS brushes bound in cold water (298 K) showed a lap shear adhesion strength lower than 0.1 MPa. The temperature dependence of the adhesion strength is related to the upper critical solution temperature (UCST) of PMAPS in water.

Since PMAPS exhibits continuous phase separation in water at UCST, it dissolves in hot water, but is insoluble in cold water at low temperature. In general, the UCST of PMAPS in water is approximately 298–303 K [52], which varies with the molecular weight of the polymer, polymer concentration, and salt concentration [53]. Recently, neutron reflectivity measurements revealed that the PMAPS brush forms a swollen structure in water above the UCST but a shrunk chain structure below room temperature [51]. A swollen and relatively extended chain structure would promote the intermixing of the opposing polymer brushes and increase the number of sulfobetaine pairs with a sufficiently close distance so that the intermolecular interaction proceeds at the brush/brush interface. The resulting substrates thus show a large adhesion strength after the removal of water and air drying.

The thermo-sensitive nature of the dipole–dipole interaction between the sulfobetaine units can be applied to adhesion strength control by solution temperature. In fact, two substrates with PMAPS brushes were smoothly debonded in water at 333 K (above the UCST), and adhered again by compression in deionized hot water and

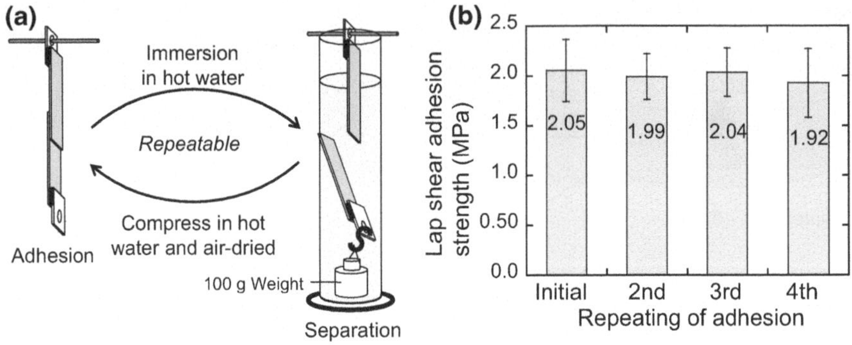

Fig. 12.4 The lap shear adhesion strength of the rebonded Si wafers with the PMAPS brushes. One adhesion cycle included adhesion in water at 333 K and air drying at 298 K for 3 h followed by the debonding process in water at 333 K

successive air drying. As shown in Fig. 12.4, the adhesive strength of the rebonded substrates was almost constant at ~2.0 MPa over repeated cycles of bonding by interfacial contact in water and separation in hot water at 333 K. These repeatable adhesion and debonding processes can be developed into novel adhesive or self-healing systems using environmentally friendly aqueous solvents.

12.4 Friction and Lubrication

The reduction in the friction coefficient under wet conditions by tethering hydrophilic polymers or polyelectrolytes on the surface has been widely accepted experimentally and theoretically. In principle, densely grafted polymer chains in a good solvent stretch from the surface to reduce their interaction and avoid overlapping with other chains. The state of chain stretching is determined by the balance between the osmotic pressure due to high polymer concentration and the elastic restoring force of the polymer chain. When two opposing, polymer brush-covered surfaces are brought into contact in a good solvent, they normally repel each other because of the excluded volume effect among polymer segments and this can suppress the mutual interpenetration of the two compressed brushes. This is the classical lubrication mechanism for efficient lubrication of solvated polymer brushes based on repulsive steric forces [54]. The reduction of frictional forces between solid surfaces bearing polymer brushes was first reported by Klein and coworkers [55] using a surface force balance. They also reported that polyelectrolyte brushes could act as efficient lubricants between mica surfaces in an aqueous solution [56–58]. Recently, extremely low friction coefficients of high-density (concentrated) poly(methyl methacrylate) brushes in toluene have been demonstrated by Tsujii et al. [59].

In the case of polyelectrolyte brushes in aqueous media, the osmotic pressure of free counterions within the charged brush also contributes to the extremely low friction property [58, 60–62]. The hydration layer bound to the charges and the fluidity of the hydrating water also play important roles in boundary lubrication [63, 64]. In addition, the lubrication properties of polyelectrolyte brushes are affected by many factors, such as graft density [65–67], ionic strength [68], solvent quality [69, 70], and the repulsive and attractive interactions of polar functional groups [71, 72], which are directly measured by surface force balance [73, 74] and AFM [75].

These studies imply that high performance of the superhydrophilic polymer brushes would be expected as a low-friction boundary layer in a water lubrication system. In particular, when two surfaces covered with ionic polymer brushes having like charges are brought into contact, the Coulombic repulsion is expected to reduce the interfacial frictional forces under wet conditions.

Takahara et al. have been investigating the macroscopic tribological characteristics of ionic [28, 76, 77] and nonionic [78, 79] polymer brushes in solvents by using a conventional ball-on-plate-type reciprocating tribotester with a glass ball probe (diameter = 10 mm) sliding on the substrates along a distance of 20 mm at a rate of 1.5×10^{-3} m/s in air and in water under a normal load of 0.49 N at 298 K, as shown in Fig. 12.5a [80, 81]. The friction coefficients at each sliding velocity were measured on a virgin surface area of the brush and silicon substrate. The root mean square (RMS) surface roughness of the glass ball was approximately 2.4 nm in a $10 \times 10\ \mu m^2$ area. The friction coefficient was determined by a strain gage attached to the arm of the tester and was recorded as a function of time. In the case of a non-modified silicon wafer under a normal load of 50 g (0.49 N), the theoretical contact area between the glass probe and substrate can be calculated as 3.51×10^{-9} m^2 by Hertz's contact mechanics theory and the average pressure on the contact area was estimated to be approximately 140 MPa.

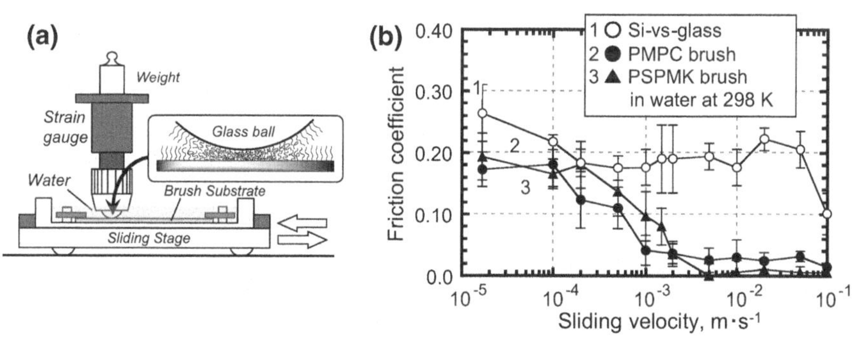

Fig. 12.5 **a** Setup of reciprocating ball-on-plate-type tribotester equipped with a water receiver on sliding stage, and **b** sliding velocity dependence of the friction coefficient of (1) non-modified silicon substrate, (2) PMPC brush, and (3) PSPMK brush in water at 298 K by sliding a glass ball (10 mm diameter) immobilized with the corresponding polymer brushes over a distance of 20 mm under a load of 50 g (0.49 N) at 298 K. A non-modified glass ball was used for (1) as a sliding probe

The friction of the glass ball covered with PSPMK or PMPC brushes sliding on the silicon substrates bearing identical brushes under a dried air atmosphere gave a friction coefficient much higher than 0.2–0.4. On the other hand, both brushes revealed a relatively lower friction coefficient than 0.2 in water. The opposing swollen brushes in water would form a thicker boundary layer to restrict the direct contact of a glass probe with the silicon substrate. Figure 12.5b shows the friction coefficients of PSPMK and PMPC brushes in water at a rate of 10^{-5}–10^{-1} m/s. The friction coefficients of these brushes in water were 0.1–0.2 at the slower friction velocity of 10^{-5}–10^{-3} m/s, whereas the friction coefficient was dropped to 0.01–0.03 at a higher sliding velocity of over 10^{-2}–10^{-3} m/s. The drastic reduction in the friction coefficients at a certain velocity could be caused by the transition of the friction mode. In other words, with the low sliding rate, the interaction between the opposite brushes and the interpenetration of brushes dominated the friction to give a large friction coefficient (boundary friction). With an increase in the sliding velocity, a thicker fluid layer would be formed between the sliding surfaces by the hydrodynamic lubrication effect to reduce the actual contact area and the friction force (mixed lubrication region). Analogous to the Stribeck curve [82], decreasing friction with increasing rate suggests that this system is in the mixed lubrication region. At higher sliding rates, hydrodynamic lubrication partially took place between swollen polyelectrolyte brushes to reduce the friction [83].

It is interesting that a transition of the lubrication mode of the PSPMK brush in water took place at the sliding velocity of 10^{-3}–10^{-1} m/s, which is relatively slower than conventional oil-based lubrication [84]. As shown in Fig. 12.5b, reduction in the friction coefficient of the non-modified silicon wafer was not observed until a velocity of 10^{-1} m/s. In other words, hydrodynamic lubrication was achieved even at a slow sliding velocity due to the polyelectrolyte brushes. This is a useful property for biological surfaces, such as synovial joints and artificial joints.

PMPC [85] is widely known as a unique polyzwitterion having excellent hydrophilicity, ionic strength dependency [86], biocompatibility [87], and nonspecific protein adsorption, due to a phosphorylcholine unit. Klein et al. also found that the PMPC brush surface under wet conditions exhibited a significantly low friction coefficient, as low as 0.0004 at pressures as high as 7.5 MPa, due to the strong hydration of the phosphorylcholine units of the polymer [88, 89]. Such a low friction performance of a PMPC brush under wet conditions has been applied to the sliding surface of an artificial hip joint [90, 91]. Lubrication in water is also expected to be promising for biolubrication of guide wires, an environmentally friendly lubrication system for rotating parts of turbines.

Wear resistance of the brush has also attracted much attention due to its practical applications. In principle, the tethered polymer chain end is bound to the substrate by a covalent bond and multiple hydrogen bonds, so that the brush layer cannot be scratched off or torn away by the sliding probe. In fact, the PSPMK brush in water under a load of 0.49 N maintained a significantly low friction coefficient continuously around 0.01 even after 400 cycles of macroscopic reciprocating friction, as shown in Fig. 12.6a. The abrasion of the brush was prevented owing to both good affinity of the PSPMK brush for water forming a water lubrication layer and the electrostatic

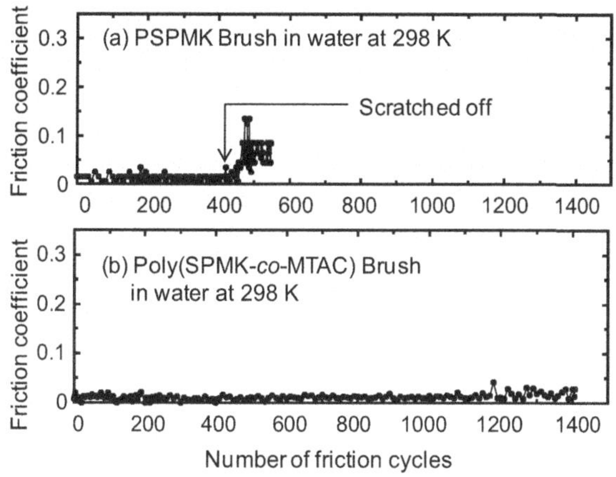

Fig. 12.6 Evolution of the friction coefficient versus number of friction cycles for the surface of **a** PSPMK and **b** poly(SPMK-*co*-MTAC) (SPMK/MTAC = 90/10) brushes by sliding a glass ball immobilized with the corresponding polymer brushes at a reciprocating distance of 10 mm and a rate of 1.5×10^{-3} m/s under a load of 0.49 N in water at 298 K

repulsive interactions among the brushes bearing sulfonic acid groups. However, the brush was eventually worn out after 450 cycles of friction, increasing the friction coefficient due to the exposure of the bare substrate.

Interestingly, the lifetime of the low friction performance was further improved by using a random copolymer brush of poly(SPMK-*co*-MTAC) [92], in which the unit molar ratio of SPMK and MTAC was 90/10, as shown in Fig. 12.6b. This ionic copolymer exhibited an extremely low friction coefficient in water of around 0.015 up to 1400 friction cycles even under a load of 0.49 N. The poly(SPMK-*co*-MTAC) brush can be regarded as a partially cross-linked polymer brush because the electrostatic attractive interaction between the sulfonate group in the SPMK unit and the ammonium group in the MTAC unit would form a cross-linking point binding the neighboring brush chains. The cross-linked structure improved the shear strength of the polymer brush preventing the wear of the brush and maintaining water in the brush layer. As a result, low friction based on water lubrication is achieved over 1000 friction cycles for a long period even under severe pressure.

12.5 Conclusions

This chapter described remarkable surface functions, such as excellent wettability, antifouling behavior repelling a hydrocarbon oil in water, reversible nanoscale adhesion without organic solvents, and significantly low friction coefficients based on water lubrication systems, achieved by high-density ion-containing polymer brushes prepared by surface-initiated controlled radical polymerization. These reversible adhesion control and water lubrication systems using polyelectrolyte brushes can contribute to innovations in novel environmentally friendly technologies. However, there are still issues remaining to understand the precise mechanism for these functions at brush interfaces. Further investigations are required to establish the relationship between the structure and function of soft interfaces by using well-designed brushes and spectroscopic analysis of the interfacial structure, in some cases combined with quantum beam measurements.

References

1. Rühe J (2004) Polymer brushes on the way to tailor-made surfaces. In: Advincula RC, Brittain WJ, Caster KC, Rühe J (eds) Polymer brushes: synthesis, characterization, applications. Wiley VCH, Weinheim, pp 1–31
2. Tsujii Y, Ohno K, Yamamoto S, Goto A, Fukuda T (2006) Structure and properties of high-density polymer brushes prepared by surface-initiated living radical polymerization. Adv Polym Sci 197:1–45
3. Zhao B, Brittain WJ (2000) Polymer brushes: surface-immobilized macromolecules. Prog Polym Sci 25:677–710
4. Kobayashi M, Ishikwa T, Takahara A (2013) Adhesion and tribological characteristics of ion-containing polymer brushes prepared by controlled radical polymerization. In: Zeng H (ed) Polymer adhesion, friction, and lubrication. John Wiley & Sons Inc., Hoboken, pp 59–82
5. Sommer S, Ekin A, Webster DC, Stafslien SJ, Daniels J, VanderWal LJ, Thompson SE, Callow ME, Callow JA (2010) A preliminary study on the properties and fouling-release performance of siloxane-polyurethane coatings prepared from poly(dimethylsiloxane) (PDMS) macromers. Biofouling 26:961–972
6. Krishnan S, Wang N, Ober CK, Finlay JA, Callow ME, Callow JA, Hexemer A, Sohn KE, Kramer EJ, Fischer DA (2006) Comparison of the fouling release properties of hydrophobic fluorinated and hydrophilic PEGylated block copolymer surfaces: attachment strength of the diatom navicula and the green alga ulva. Biomacromol 7:1449–1462
7. Yarbrough JC, Rolland JP, DeSimone JM, Callow ME, Finlay JA, Callow JA (2006) Contact angle analysis, surface dynamics, and biofouling characteristics of cross-linkable, random perfluoropolyether-based graft terpolymers. Macromolecules 39:2521–2528
8. Brady RF Jr, Singer IL (2000) Mechanical factors favoring release from fouling release coatings. Biofouling 15:73–81
9. Lejars M, Margaillan A, Bressy C (2012) Fouling release coatings: a nontoxic alternative to biocidal antifouling coatings. Chem Rev 112:4347–4390
10. Krishnan S, Weinman CJ, Ober CK (2008) Advances in polymers for anti-biofouling surfaces. J Mater Chem 18:3405–3413
11. Dalsin JL, Messersmith PB (2005) Bioinspired antifouling polymers. Mater Today 8:38–46
12. Hucknall A, Rangarajan S, Chilkoti A (2009) In pursuit of zero: polymer brushes that resist the adsorption of proteins. Adv Mater 21:2441–2446
13. Tripathi BP, Dubey NC, Choudhury S, Stamm M (2012) Antifouling and tunable amino functionalized porous membranes for filtration applications. J Mater Chem 22:19981–19992
14. Li GZ, Cheng G, Xue H, Chen S, Zhang F, Jiang S (2008) Ultra low fouling zwitterionic polymers with a biomimetic adhesive group. Biomaterials 29:4592–4597
15. Higaki Y, Nishida J, Takenaka A, Yoshimatsu R, Kobayashi M, Takahara A (2015) Versatile inhibition of marine organism settlement by zwitterionic polymer brushes. Polym J 47:811–818
16. Holmlin RE, Chen X, Chapman RG, Takayama S, Whitesides GM (2001) Zwitterionic SAMs that resist nonspecific adsorption of protein from aqueous buffer. Langmuir 17:2841–2850

17. Jiang S, Cao Z (2010) Ultralow-fouling, functionalizable, and hydrolyzable zwitterionic materials and their derivatives for biological applications. Adv Mater 22:920–932
18. Ederth T, Ekblad T, Pettitt ME, Conlan SL, Du CX, Callow ME, Callow JA, Mutton R, Clare AS, D'Souza F, Donnelly G, Bruin A, Willemsen PR, Su XJ, Wang S, Zhao Q, Hederos M, Konradsson P, Liedberg B (2011) Resistance of galactoside-terminated alkanethiol self-assembled monolayers to marine fouling organisms. ACS Appl Mater Interfaces 3:3890–3901
19. Luk YY, Kato M, Mrksich M (2000) Self-assembled monolayers of alkanethiolates presenting mannitol groups are inert to protein adsorption and cell attachment. Langmuir 16:9604–9608
20. Chelmowski R, Köster SD, Kerstan A, Prekelt A, Grunwald C, Winkler T, Metzler-Nolte N, Terfort A, Wöll C (2008) Peptide-based SAMs that resist the adsorption of proteins. J Am Chem Soc 130:14952–14953
21. Statz AR, Meagher RJ, Barron AE, Messersmith PB (2005) New peptidomimetic polymers for antifouling surfaces. J Am Chem Soc 127:7972–7973
22. Ostuni E, Chapman RG, Holmlin RE, Takayama S, Whitesides GM (2001) A survey of structure-property relationships of surfaces that resist the adsorption of protein. Langmuir 17:5605–5620
23. Kobayashi M, Terayama Y, Yamaguchi H, Terada M, Murakami D, Ishihara K, Takahara A (2012) Wettability and antifouling behavior on the surfaces of superhydrophilic polymer brushes. Langmuir 28:7212–7222
24. Murakami D, Kobayashi M, Moriwaki T, Ikemoto Y, Jinnai H, Takahara A (2013) Spreading and structuring of water on superhydrophilic polyelectrolyte brush surfaces. Langmuir 29:1148–1151
25. Yamaguchi H, Honda K, Kobayashi M, Morita M, Masunaga H, Sakata O, Sasaki S, Takahara A (2008) Molecular aggregation state of surface-grafted poly{2-(perfluorooctyl)ethyl acrylate} thin film analyzed by grazing incidence x-ray diffraction. Polym J 40:854–860
26. Yamaguchi H, Kikuchi M, Kobayashi M, Ogawa H, Masunaga H, Sakata O, Takahara A (2012) Influence of molecular weight dispersity of poly{2-(perfluorooctyl)ethyl acrylate} brushes on their molecular aggregation states and wetting behavior. Macromolecules 45:1509–1516
27. Kobayashi M, Terada M, Terayama Y, Kikuchi M, Takahara A (2010) Direct synthesis of well-defined poly[{2-(methacryloyloxy)ethyl}trimethylammonium chloride] brush via surface-initiated atom transfer radical polymerization in fluoroalcohol. Macromolecules 43:8408–8415
28. Kobayashi M, Terayama Y, Hosaka N, Kaido M, Suzuki A, Yamada N, Torikai N, Ishihara K, Takahara A (2007) Friction behavior of high-density poly(2-methacryloyloxyethyl phosphorylcholine) brush in aqueous media. Soft Matter 3:740–746
29. Owens DK, Wendt RC (1969) Estimation of the surface free energy of polymers. J Appl Polym Sci 13:1741–1747
30. Kobayashi M, Matsugi T, Saito J, Imuta J, Kashiwa N, Takahara A (2013) Direct modification of polyolefin films by surfaceinitiated polymerization of a phosphobetaine monomer. Polym Chem 4:731–739
31. Chen KS, Uyama Y, Ikada Y (1994) Adhesive interaction between polymer surfaces grafted with water soluble polymer chains. Langmuir 10:1319–1322
32. Thünemann AF, Müller M, Dautzenberg H, Joanny JF, Löwen H (2004) Polyelectrolyte complexes. Adv Polym Sci 166:113–171
33. LaSpina R, Tomlinson MR, Ruiz-Pérez L, Chiche A, Langridge S, Geoghegan M (2007) Controlling netweork-brush interactions to achieve switchable adhesion. Angew Chem Int Ed 46:6460–6463
34. Sudre G, Olanier L, Tran Y, Hourdet D, Creton C (2012) Reversible adhesion between a hydrogel and a polymer brush. Soft Matter 8:8184–8193
35. Kobayashi M, Terada M, Takahara A (2011) Reversible adhesive-free nanoscale adhesion utilizing oppositely charged polyelectrolyte brushes. Soft Matter 7:5717–5722
36. Li ZF, Kang ET, Neoh KG, Tan KL, Huang CC, Liaw DJ (1997) Surface structures and adhesive-free adhesion characteristics of polyaniline films after modification by graft copolymerization. Macromolecules 30:3354–3362

37. Kang ET, Shi JL, Neoh KG, Tan KL, Liaw DJ (1998) Surface modification of polytetrafluoroethylene films via graft copolymerization for auto-adhesion. J Polym Sci Part A Polym Chem 36:3107–3114
38. Han HS, Tan KL, Kang ET, Neoh KG (1998) Low-temperature graft copolymerization of 1-vinyl imidazole on low-density polyethylene films with simultaneous lamination of copper foils. J Appl Polym Sci 70:1977–1983
39. Wang T, Kang ET, Neoh KG, Tan KL, Liaw DJ (1998) Surface modification of low-density polyethylene films by UV-induced graft copolymerization and its relevance to photolamination. Langmuir 14:921–927
40. Chen W, Neoh KG, Kang ET, Tan KL, Liaw DJ, Huang CC (1998) Surface modification and adhesion characteristics of polycarbonate films after graft copolymerization. J Polym Sci Part A Polym Chem 36:357–366
41. Kang ET, Neoh KG, Li ZF, Tan KL, Liaw DJ (1998) Surface modification of polymer films by graft copolymerization for adhesive-free adhesion. Polymer 39:2429–2436
42. Ma ZH, Han HS, Tan KL, Kang ET, Neoh KG (1999) Surface graft copolymerization induced adhesion of polyaniline film to polytetra-fluoroethylene film and copper foil. Eur Polym J 35:1279–1288
43. Liu YX, Kang ET, Neoh KG, Tan KL, Huang CC, Liaw DJ (1999) Lamination of polytetrafluoroethylene films via surface thermal graft copolymerization with ionic and zwitterionic monomers. J Appl Polym Sci 74:816–824
44. Ma ZH, Han HS, Tan KL, Kang ET, Neoh KG (1999) Thermally induced surface graft copolymerization with concurrent lamination of polyaniline films under atmospheric conditions. Int J Adhes Adhes 19:359–365
45. Lowe AB, McCormick CL (2002) Synthesis and solution properties of zwitterionic polymers. Chem Rev 102:4177–4189
46. Lowe AB, McCormick CL (2006) Synthesis aqueous solution properties, and biomedical application of polymeric betaines. In: Lowe AB, McCormick CL (eds) Polyelectrolytes and polyzwitterions: synthesis, properties, and applications. ACS Books, Washington DC, pp 65–78
47. Chen S, Jiang S (2008) An new avenue to nonfouling materials. Adv Mater 20:335–338
48. West SL, Salvage JP, Lobb EJ, Armes SP, Billingham NC, Lewis AL, Hanlon GW, Lloyd AW (2004) The biocompatibility of crosslinkable copolymer coatings containing sulfobetaines and phosphobetaines. Biomaterials 25:1195–1204
49. Zhang Z, Zhang M, Chen S, Horbett TA, Ratner BD, Jiang S (2008) The hydrolysis of cationic polycarboxybetaine esters to zwitterionic polycarboxybetaines with controlled properties. Biomaterials 29:4719–4725
50. Ladd J, Zhang Z, Chen S, Hower JC, Jiang S (2008) Zwitterionic polymers exhibiting high resistance to nonspecific protein adsorption from human serum and plasma. Biomacromol 9:1357–1361
51. Kobayashi M, Takahara A (2013) Environmentally friendly repeatable adhesion using a sulfobetaine-type polyzwitterion brush. Polym Chem 4:4987–4992
52. Schulz DN, Peiffer DG, Agarwal PK, Larabee J, Kaladas JJ, Soni L, Handwerker B, Garner RT (1986) Phase behaviour and solution properties of sulphobetaine polymers. Polymer 27:1734–1742
53. Chen L, Honma Y, Mizutani T, Liaw DJ, Gong JP, Osada Y (2000) Effects of polyelectrolyte complexation on the UCST of zwitterionic polymer. Polymer 41:141–147
54. Klein J (2001) Interactions friction and lubrication between polymer-bearing surfaces In: Bhushan B (ed) Fundamentals of tribology and bridging the gap between the macro- and micro/nanoscales. Kluwer Academic Publishers, Dordrecht, pp 177–198
55. Klein J, Perahia D, Warburg S (1991) Forces between polymer-bearing surfaces undergoing shear. Nature 352:143–145
56. Klein J, Kumacheva E, Mahalu D, Perahia D, Fetters LJ (1994) Reduction of frictional forces between solid surfaces bearing polymer brushes. Nature 370:634–636
57. Kampf N, Gohy JF, Jerome R, Klein J (2005) Normal and shear forces between a polyelectrolyte brush and a solid surface. J Polym Sci Part B Polym Phys 43:193–204

58. Raviv U, Giasson S, Kamph N, Gohy JF, Jérôme R, Klein J (2003) Lubrication by charged polymers. Nature 425:163–165
59. Nomura A, Okayasu K, Ohono K, Fukuda T, Tsujii Y (2011) Lubrication mechanism of concentrated polymer brushes in solvents: effect of solvent quality and thereby swelling state. Macromolecules 44:5013–5019
60. Ruhe J, Ballauff M, Biesalski M, Dziezok P, Gröhn F, Johannsmann D, Houbenov N, Hugenberg N, Konradi R, Minko S, Motornov M, Netz RR, Schmidt M, Seidel C, Stamm M, Stephan T, Usov D, Zhang H (2004) Adv Polym Sci 165–166:79–150
61. Luzinov I, Minko S, Tsukruk VV (2004) Adaptive and responsive surfaces through controlled reorganization of interfacial polymer layers. Prog Polym Sci 29:635–698
62. Miklavic SJ, Marčelja S (1988) Interaction of surfaces carrying grafted polyelectrolytes. J Phys Chem 92:6718–6722
63. Klein J, Raviv U, Perkin S, Kampf N, Chai L, Giasson S (2004) Fluidity of water and of hydrated ions confined between solid surfaces to molecularly thin films. J Phys Condens Matter 16:S5437–S5448
64. Raviv U, Klein J (2002) Fluidity of bound hydration layers. Science 297:1540–1543
65. Israels R, Leermakers FAM, Fleer GJ, Zhulina EB (1994) Charged polymeric brushes: structure and scaling relations. Macromolecules 27:3249–3261
66. Lyatskaya YV, Leermakers FAM, Fleer GJ, Zhulina EB, Birshtein TM (1995) Analytical self-consistent-field model of weak polyacid brushes. Macromolecules 28:3562–3569
67. Zhulina EB, Wolterink JK, Borisov OV (2000) Screening effects in a polyelectrolyte brush: self-consistent-field theory. Macromolecules 33:4945–4953
68. Pincus P (1991) Colloid stabilization with grafted polyelectrolytes. Macromolecules 24:2912–2919
69. Ross RS, Pincus P (1992) The polyelectrolyte brush: poor solvent. Macromolecules 25:2177–2183
70. Pryamitsyn VA, Leermakers FAM, Fleer GJ, Zhulina EB (1996) Theory of the collapse of the polyelectrolyte brush. Macromolecules 29:8260–8270
71. Taunton HJ, Toprakcioglu C, Fetters LJ, Klein J (1988) Forcers between surfaces bearing terminally anchored polymer chains in good solvents. Nature 332:712–714
72. Eiser E, Klein J, Witten TA, Fetters LJ (1999) Shear of telechelic brushes. Phys Rev Lett 82:5076–5079
73. Hayashi S, Abe T, Higashi N, Niwa M, Kurihara K (2002) Polyelectrolyte brush layers studied by surface forces measurement: dependence on ph and salt concentrations and scaling. Langmuir 18:3932–3944
74. Kampf N, Ben-Yaakov D, Andelman D, Safran SA, Klein J (2009) Direct measurement of sub-debye-length attraction between oppositely charged surfaces. Phys Rev Lett 103:118304
75. Kelley TW, Shorr PA, Kristin DJ, Tirrell M, Frisbie CD (1998) Direct force measurements at polymer brush surfaces by atomic force microscopy. Macromolecules 31:4297–4300
76. Kobayashi M, Takahara A (2010) Tribological properties of hydrophilic polymer brushes under wet conditions. Chem Record 10:208–216
77. Ishikawa T, Kobayashi M, Takahara A (2010) Macroscopic frictional properties of poly(1-(2-methacryloyloxy)ethyl-3-butyl Imidazolium Bis(trifluoromethanesulfonyl)imide) brush surfaces in an ionic liquid. Appl Mat Interfaces 2:1120–1128
78. Sakata H, Kobayashi M, Otsuka H, Takahara A (2005) Tribological properties of poly(methyl methacrylate) brushes prepared by surface-initiated atom transfer radical polymerization. Polym J 37:767–775
79. Kobayashi M, Takahara A (2005) Synthesis and frictional properties of poly(2,3-dihydroxypropyl methacrylate) brush prepared by surface-initiated atom transfer radical polymerization. Chem Lett 34:1582–1583
80. Kobayashi M, Terada M, Ishikawa T, Takahara A (2013) Tribological behavior of ionic polymer brushes in aqueous environment. In: Biresaw G, Mittal KL (eds) Surfactants in tribology, vol 3. CRC Press, Boca Raton, pp 75–92

81. Kobayashi M, Wang Z, Matsuda Y, Kaido M, Suzuki A, Takahara A (2009) Tribological behavior of polymer brush prepared by the "grafting-from" method. In: Kumar SS (ed) Polymer Tribology. Imperial College Press, UK, pp 582–602
82. Spikes HA (1993) Boundary lubrication and boundary films. In: Dowson D, Talor CM, Childs THC, Godet M, Dalmz G (eds) Thin films in tribology. Elsevier, New York, pp 331–346
83. Kobayashi M, Tanaka H, Minn M, Sugimura J, Takahara A (2014) Interferometry study of aqueous lubrication on the surface of polyelectrolyte brush. ACS Appl Mater Interfaces 6: 20365–20371
84. Bielecki RM, Crobu M, Spencer ND (2013) Polymer-Brush lubrication in oil: sliding beyond the stribeck curve. Tribol Lett 49:263–272
85. Ishihara K, Ueda T, Nakabayashi N (1990) Preparation of phospholipid polymers and their properties as polymer hydrogel membranes. Polym J 22:355–360
86. Matsuda Y, Kobayashi M, Annaka M, Ishihara K, Takahara A (2008) Dimensions of a free linear polymer and polymer immobilized on silica nanoparticles of a zwitterionic polymer in aqueous solutions with various ionic strengths. Langmuir 24:8772–8778
87. Ishihara K, Nomura H, Mihara T, Kurita K, Iwasaki Y, Nakabayashi N (1998) Why do phospholipid polymers reduce protein adsorption? J Biomed Mater Res 39:323–330
88. Chen M, Briscoe WH, Armes SP, Klein J (2009) Lubrication at physiological pressures by polyzwitterionic brushes. Science 323:1698–1701
89. Chen M, Briscoe WH, Armes SP, Cohen H, Klein J (2011) Polyzwitterionic brushes: extreme lubrication by design. Eur Polym J 47:511–523
90. Moro T, Takatori Y, Ishihara K, Konno T, Takigawa Y, Matsushita T, Chung U, Nakamura K, Kawaguchi H (2004) Surface grafting of artificial joints with a biocompatible polymer for preventing periprosthetic osteolysis. Nature Mater 3:829–836
91. Moro T, Kawaguchi H, Ishihara K, Kyomoto M, Karita T, Ito H, Nakamura K, Takatori Y (2009) Wear resistance of artificial hip joints with poly(2-methacryloyloxyethyl phosphorylcholine) grafted polyethylene: Comparisons with the effect of polyethylene cross-linking and ceramic femoral heads. Biomaterials 30:2995–3001
92. Kobayashi M, Takahara A (2012) Polyelectrolyte brushes: a novel stable lubrication system in aqueous conditions. Faraday Discuss 156:403–412

Chapter 13
Bio- and Chemical Sensors and Role of Soft Interface

Yukari Sato

13.1 Bio- and Chemical Sensors and Soft Interfaces (Introduction)

A soft interface and a solid surface modified with soft materials are of great interest as regards the construction of an intelligent interface for assembling bio-/chemical sensors, because an interface modified with soft materials can be used to provide solid materials with various new functions. An interface modified with soft material has been employed to fabricate chemical/biochemical sensors. One typical example of a biosensing system that uses a soft interface is the glucose sensor. Glucose sensor systems are now portable, very small, and inexpensive. Many companies have marketed their own glucose-sensing systems. A glucose sensor has many components, but one of the most important technologies involves the fabrication of an enzyme-attached electrode surface covered with intelligent polymers, for example, mediators and enzymes. The method for fabricating an enzyme-attached electrode has become a key technology. Blood samples contain not only blood glucose but also a very large volume of blood corpuscles, proteins, lipids, organic/inorganic substances, and other components. It is difficult to detect only glucose clearly.

Typically, bio- or chemical-sensing systems are constructed with the specific items as shown in Fig. 13.1. Target molecules (biomolecules) are recognized and concentrated on the soft interface modified surface with specific affinities for target molecules (such as enzymes, antibodies, chemicals, other molecules, and devices). After detecting specific molecules, mass changes, refractive index changes, conductivity changes, and other responses occur and are recorded. The variation is displayed simply or translated to other signals. With a very simple system such as one involving antigen–antibody affinity caused by monitored surface mass changes or refrac-

Y. Sato (✉)
National Institute of Advanced Industrial Science and Technology (AIST), 1-1-1 Umezono, Tsukuba, Ibaraki 305-8560, Japan
e-mail: yukari-sato@aist.go.jp

M. Maeda et al. (eds.), *Molecular Soft-Interface Science*,
https://doi.org/10.1007/978-4-431-56877-3_13

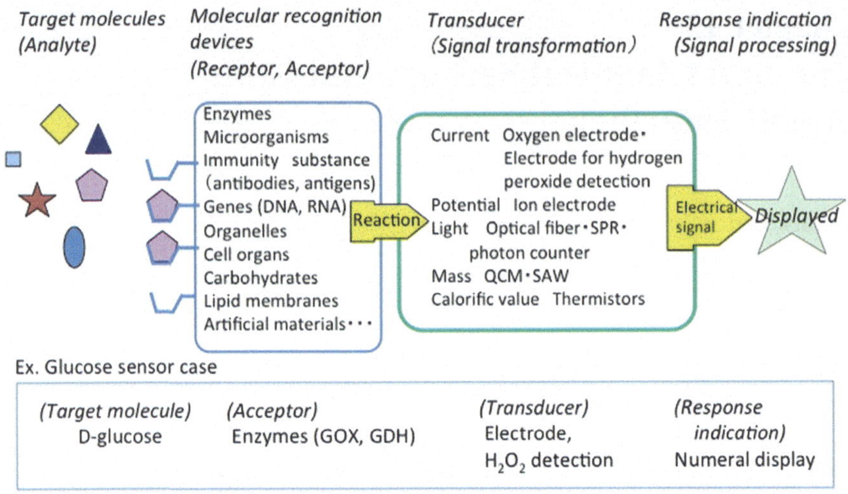

Fig. 13.1 Scheme of bio- and chemical- sensing systems (analyte, receptor, transducer, and signal processing). The example relates to glucose sensor in the bottom

tive index changes, the result (the degree of recognition) is displayed directly and immediately. But above all, bio- and chemical-sensing need effective signal transformation and transducers. When the affinity response is very small, it must be boosted significantly using an amplifier, and other modifications must be made including the provision of tags, markers, and weights. Quartz crystal microbalance (QCM) measurement is a simple sensing method for evaluating bio- or chemical affinity from surface mass changes. Electrochemical measurement is another way of detecting chemical and biochemical interactions. With this approach, the dynamic changes of oxidative/reductive currents and electric charges can be monitored. When electrochemical active tags are used, the electrochemical-activated materials generated after enzyme/chemical reactions and the fluctuation of the electrochemically activated materials in the measurement system make electrochemical detection effective. Variations in the impedance of the recognition surface are also used to evaluate the molecular sensing. Finally, the way the results are displayed is also important. The signal response should be indicated clearly and immediately once the target molecules have been recognized.

With the abovementioned glucose sensor, the target molecule is D-glucose and the molecular acceptor/ligand is an enzyme (glucose oxidase or glucose dehydrogenase). The signal transformation is realized with the electrochemical method (the current used for hydrogen peroxide detection is measured) (Fig. 13.1). To obtain accurate data (glucose concentration) from a real blood sample, the electrode surface was covered with intelligent polymers thus avoiding the nonspecific adsorption of extraneous proteins and chemicals. The method for anchoring the enzyme (glucose oxidase or glucose dehydrogenase) on the electrode surface also becomes important. To avoid denaturing the enzymes when the enzyme is firmly and directly attached to the

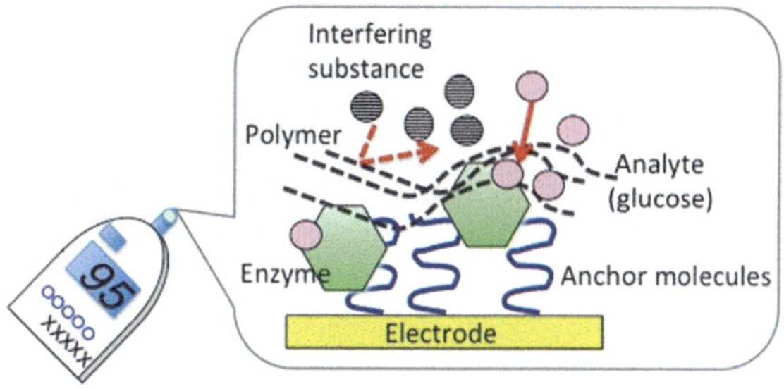

Fig. 13.2 Image of the biosensor and the role of the soft interface for the glucose sensor

electrode surface, the enzyme could be soft landed on the electrode surface using polymers or other certain materials within mediators for electron transfer. A sensor electrode that operates for a long time is also needed. The modified electrode and other substrates must be disposable because blood samples are handled. The total cost must be reduced, so each component is constructed from inexpensive materials (Fig. 13.2).

When an interaction between biomolecules such as antigen–antibody or enzyme-specific substrates is detected, the detection chip surface and its modification become very important. The stable immobilization of ligand materials on the chip surface and the avoidance of nonspecific adsorption on the recognition surface are important issues if we are to receive satisfactory data. For example, the Biacore system series (GE Healthcare), a surface plasmon resonance (SPR) detection method, and a polymer-modified gold chip surface are used for evaluating (bio) molecule interactions. Moreover, in a DNA chip and protein chip surfaces, chemical modification on the chip substrates is important in terms of achieving high performance. On the DNA chip, the DNA chain anchored on the surface and unrelated co-adsorbed molecules to prevent nonspecific adsorption are important as regards obtaining target single-strand DNAs. The unrelated molecules provide a certain space and flexibility for reaction with the true target DNA sequence. There are many factors that must be considered if we are to construct and to operate sensors. There are many tasks related to soft interfaces, soft materials, chemicals, and substrate (electrode) surfaces that must be completed if we are to construct satisfactory sensors.

Table 13.1 showed typical transducers that are often used for (bio)sensing and detecting target molecules. Table 13.1 lists physical quantities, detection methods, transducers, biosensing materials (ligands), and detection targets (analytes). With electric detection, typical biosensing methods are current detection, voltage (potential) detection, and resistance detection. An oxygen electrode is often used as a transducer for hydrogen peroxide detection and in cyclic voltammogram methods. pH, ion, and gas detection electrodes are typically adopted for potential detection. Impedance measurement is also commonly used to evaluate the interaction between biomolecules. These transducers are used for bio(chemical) sensing in combination with ligand molecules. Ligands can be either natural products or artificial compounds. Enzymes, microorganisms, and lipid membranes are often used. Analytes include glucose, lactic acid, glutamic acid, and potassium ions. With optical detec-

Table 13.1 Typical transducers and detection targets

Physical quantities	Detection method	Transducers (e.g.)	Ligand (e.g.)	Analyte (e.g.)
Electric/electricity	Current Potential/voltage Resistance	Oxygen electrode H_2O_2 electrode cyclic voltammogram pH electrodeion electrode Gas electrode impedance	Enzyme, microbe, organs Enzyme Enzyme Enzyme Enzyme, microbe, Ionophore Enzyme Ionophore, Lipid membrane	Glucose, glutamate, lactate, BOD Glucose, lactate, pyruvate creatine, NAD+ H^+, urea, insecticide glucose, glutamate,K^+,NH_3 glucose, urea K +, acetylcholine, glutamate
Light/optical	Fluorescence Luminescence Adsorption Surface plasmon Resonance	Optical fiber Photon Counter Optical fiber CCD pixel	Enzyme, protein Enzyme, antibody Hemoglobin Antibody, biotin	H^+, albumin H_2O_2 antigen oxygen Antigen, avidin
Calorie/quantity of heat	Calorie	Thermistor capacitive measurement	Enzyme Enzyme	Glucose, urea, ethanol toxin
Mass	Frequency change	Quartz crystal microbalance (QCM), surface acoustic wave device (SAW)	Antibody	Antigen, chemicals

tion (amounts of light, photons, and luminous energy), photon counters and optical fibers are used as transducers, lectin (proteins), enzymes, antibodies, and other substances are adopted as biosensing materials (ligands). Oxygen, hydrogen peroxide (chemicals), and antigens (biomaterials) are detected as analytes. Moreover, calorie and mass measurements are also used. For transducers, heat capacity measurements, thermistors (for calorie measurement), quartz crystal microbalance (QCM) measurements, and surface acoustic wave (SAW) devices (for mass measurement) are also used. The ligands are enzymes and antibodies, and the analytes are glucose, urea, antigen, toxins, and chemicals (agricultural chemicals).

This chapter introduces and describes measurement methods that are often adopted for (bio)chemical sensors. Many smart measurement methods are available for evaluating (bio)molecular interactions. As shown in Table 13.2, simple and high detection limit measurements for bio- and chemical measurements are observed. As regards optical measurements, surface plasmon resonance (SPR) measurement is often adopted to detect the interaction between biomolecules. Mass change detection is also used to evaluate biomolecular affinities. Quartz crystal microbalance measurement (QCM) has been put to general use. In the early stages of QCM measurement for evaluating biomolecular affinities, simple mass changes on the modified

Table 13.2 Detection methods for bio-, chemical sensors: ranges of detection concentration, measurement time, convenience, and quantitatively

Detection method	Concentration (mol/L)	Measurement time	Convenience	Quantitatively
Electrochemical	$10^0 \sim 10^{-6}$	Few minutes	Excellent	Good
Immunochromatographic	$10^{-2} \sim 10^{-8}$		Excellent	Poor
SPR	$10^{-2} \sim 10^{-8}$		Good	Good
Frequency change (QCM)	$10^0 \sim 10^{-8}$		Good	Good
Mass change detection	$10^{-8} \sim 10^{-9}$		Poor	Excellent
Field-effect transistor (FET)	$10^{-8} \sim 10^{-10}$		Fair	Good
Electrochemiluminescence (ECL)	$10^{-8} \sim 10^{-13}$		Good	Good
ELISA (fluorescence)	$10^{-6} \sim 10^{-9}$		Fair	Good
Chemiluminescence	$10^{-7} \sim 10^{-10}$	Over ten hours	Good	Good
Radioimmunoassay	$10^{-7} \sim 10^{-12}$		Poor	Good

(including ligand) surface were simply monitored with and without the detection target molecules. Today, the monitoring of mass changes on a ligand surface has changed. When recognition occurs on a surface, not only real mass changes but also the viscidity and elasticity of the recognition monolayer are monitored simultaneously. Electrochemical measurement is often adopted to evaluate the affinity between biomolecules because electrochemical measurement is very sensitive to changes in surface materials.

First, we describe the methodology of the evaluation of surface materials, surface modification layers, and sensing methods for bio/chemical molecules detection. We introduce the most important, widely used, highly sensitive, and user-friendly sensing methods for surface evaluation technology, including surface plasmon resonance measurement (SPR), quartz crystal microbalance (QCM), electrochemical measurement (cyclic voltammetry (CV) and others), the field-effect transistor (FET) method, and other methods for evaluating molecular affinities and detecting limited target molecules.

13.2 Methodology of Sensing/Biosensing—Electrochemistry, QCM, SPR, and FET

13.2.1 Quartz Crystal Microbalance (QCM)

The quartz crystal microbalance (QCM) represents one of the classes of quartz oscillators capable of detecting small mass changes in surfaces [1]. The QCM consists of a piezoelectric quartz crystal disk coated with a gold (or other metal electrodes) thin film. This is coupled to an oscillating circuit that applies an alternating electrical field to the crystal. A frequency counter, which monitors the changes in the crystal frequency, is linked directly to the circuit. When a material is deposited on the crystal surface, the wave's amplitude is reduced and the resonance frequency decreases. The adsorbed mass is evaluated by using the Sauerbrey equation. The frequency variations are related to the mass deposition in Eq. (13.1) as follows:

$$\Delta f = - \left(2\Delta m f_0^2\right)/A\left(\mu_q \rho_q\right)^{1/2} \tag{13.1}$$

where ρ_q is the quartz density, μ_q is the crystal shear modulus, f_0 is the crystal's fundamental frequency, A is the crystal's piezoelectrically active geometrical area, and Δm and Δf are mass and frequency changes, respectively. Under certain conditions, a change in frequency is proportional to mass changes in the surface films on the electrodes. Equation 13.1 is adopted for gas phase studies. In the liquid phase, almost all cases can also employ Eq. 13.1. Constant value of Eq. 13.1 was compiled as C (sensitivities) and Eq. 13.1 was changed to Eq. (13.2) as given below:

$$\Delta m = - C\,\Delta f \tag{13.2}$$

With an AT-cut, 5 MHz quartz crystal microbalance, C becomes 17.7×10^{-9}g $\cdot$Hz$^{-1}\cdot$cm^{-2}. A 1 Hz change corresponds to a 17.7 ng$\cdot$cm^{-2} variation. When an AT-cut, 9 MHz quartz crystal was used, C became 4.4×10^{-9}g$\cdot$Hz$^{-1}\cdot$cm^{-2}, and so a 1 Hz frequency change corresponded to a 4.4 ng$\cdot$cm^{-2} variation on the surface. Almost all materials, even if it is a monomolecular layer, and its adsorption/desorption to the substrate surface, can be detected well. Table 13.3 shows the frequency response on the surface. It is also important to consider the noise response. A 0.1 Hz level noise was observed with a 5 MHz QCM, so we can detect the adsorption response of most of the materials on the surface, even in terms of monolayer adsorption. The expected frequency changes for monolayer adsorption on the QCM surface (ex. on the Pt electrode where 1.3×10^{15} molecules $\cdot$cm^{-2} will be detected for one Pt molecule attached) are shown in Table 13.3. With a thin monolayer, Eqs. 13.1 and 13.2 can

Table 13.3 Expected frequency changes in monolayer adsorption (for 5 MHz QCM)

Chemicals (adsorption materials)	Molar mass (g/mol)	Calculated/expected frequency changes (Hz)
H	1.0	0.12
Li	6.9	0.86
Pb	207	25.7
H_2O	18	2.2
SO_4^{2-}	96	11.9
$CH_3(CH_2)_{11}SH$	202	25.1

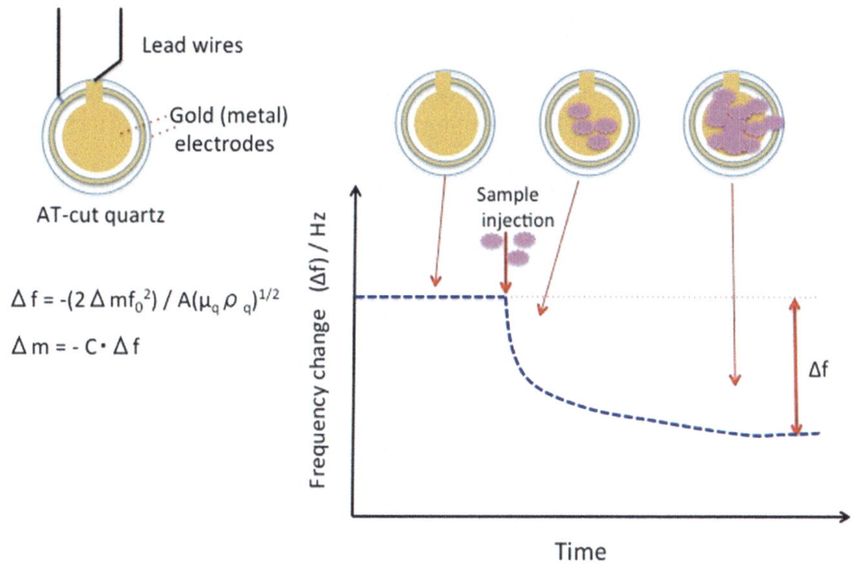

Fig. 13.3 Molecular adsorption monitoring by QCM method

be used to convert the frequency to mass changes. Adsorption monitoring on a gold surface is shown schematically in Fig. 13.3.

Unfortunately, Eq. 13.2 is not applicable to the surface adsorption of large molecular weight polymers and biomaterials with a large volume of water molecules. In such cases, resonance frequency, Δf, and energy dissipation, ΔD, are monitored simultaneously. Dissipation factor, D (dimensionless units, $\times 10^{-6}$), provides data regarding the viscoelasticity of the attached layer, giving information about the strength of attachment and the structure of the attached layer. A study of the dissipation as a function of the frequency shift (f versus D plots) can often reveal information regarding film strength, film interaction, and the identification of the various phases that the film has undergone. Analysis of the Δf and ΔD values at multiple overtones can generate information regarding the thickness, shear, and viscosity of the attached layer. Equation 13.2 was changed to Eq. 13.3 as shown below:

$$\Delta m = - C' \Delta f / n \tag{13.3}$$

where n is the overtone number (1,3,5,7…..) and C' is mass sensitivity ($17.7 \times 10^{-9} g \cdot Hz^{-1} \cdot cm^{-2}$ at an oscillation frequency of 5 MHz, overtone n = 1). The dissipation factor D is defined as

$$D = E_{Dissipated} / 2\pi E_{Stored} \tag{13.4}$$

where $E_{Dissipated}$ is the energy dissipated during one oscillation period and E_{Stored} is the energy stored during the oscillation. In contrast to rigid films, the viscoelastic properties of soft matter give rise to energy dissipation. Studies of mass change (Δm) measurements with a D value for evaluating soft material or a soft interface increased rapidly [2–6].

QCM measurement can also be combined with the electrochemical method. If one of the gold (metal) films on either side of crystal also functions as a working electrode as in an electrochemical cell, the device is known as a working electrode for the electrochemical quartz crystal microbalance (EQCM). This combination measurement can reveal mass changes on the electrochemical reactions [7–10].

13.2.2 Surface Plasmon Resonance (SPR) Measurement

SPR measurements are widely used in the detection of bio- and chemical molecules. SPR is a sensitive optical technique for the detection of several biomolecular interactions such as peptide–antibody, protein–DNA, and protein—polysaccharide interactions. The plasmon phenomenon takes places when an incident light beam with an appropriate wavelength comes into contact with a thin metal film, usually gold, at a particular angle (SPR angle). As a result, an evanescent electromagnetic field is generated on the surface of the disk. Hence, a charge density oscillation occurs at the interface between two media with oppositely charged dielectric constants. When there is an interaction between the immobilized molecule and the analyte, a shift is induced in the SPR resonance angle that is directly related to the number of molecules that interact with the sensor surface.

SPR measurements use near-field optics, which means that the detection field is limited to a few hundred nanometers from the metal surface. The basic principle and operation of SPR have been described in several reviews [11]. An evanescent field is generated under total internal reflections occurring on a transducer surface. Two kinds of configuration are often used for surface plasmon excitation, namely, the Kretschmann and Otto configurations. Most SPR instruments use a Kretschmann configuration functioning with attenuated total reflectance (ATR) for the excitation of surface plasmons. Generally, an SPR sensor is comprised of several important components, namely, a light source, a detector, a transduction surface, usually gold film, a prism, biomolecules (ex. antibody–antigen) and a flow system. Figure 13.4 shows the scheme of the principle and operation of the SPR immunoassay technique. The transduction surface is usually a thin gold film, with a thickness of 50~100 nm, on a glass slide optically coupled to a glass prism through refractive index matching oil. In addition to gold, several other metals can be used including silver, copper, and aluminum. Gold is strongly preferred due to its chemical stability and free electron behavior. Plane polarized light is directed through a glass prism to the gold/solution dielectric interface over a wide range of incident angles and the intensity of light angle is measured with a detector. At certain incident light wavelengths and angles, the reflectivity exhibits a minimum value at which the light waves are coupled to

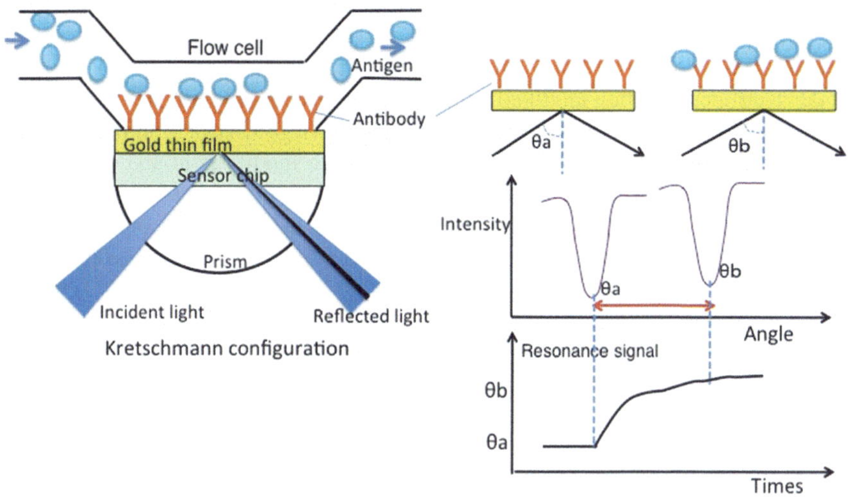

Fig. 13.4 Surface plasmon resonance (SPR) immunoassay technique

the oscillation of surface plasmons at the gold/solution interface. The angle at which the minimum reflectivity occurs is denoted as the SPR angle. This critical angle is very sensitive to the dielectric properties of the medium adjacent to the transducer surface apart from its dependence on their wavelength and the polarization state of the incident light. In particular, the resonance condition is extremely sensitive to the refractive index of the sample in contact with the metal surface because the optical electric fields are localized to within 200~250 nm of the gold surface.

The resonance conditions are influenced by the biomolecules immobilized on the gold surface. The adsorption of biomolecules on the metallic surface, followed by conformational changes in the adsorbed biomolecules and molecular interactions with relevant substances can be accurately detected. When a surface-immobilized antibody binds with an analyte, the change in the interfacial refractive index can be detected as a shift in the resonance angle. These changes provide information on the amount of bound analyte, the affinity of the analyte, and the association/dissociation kinetics in relation to the analyte. There are many companies making SPR instruments for use in investigating biomolecular interactions. Each group produces original SPR systems equipped with a variety of options for specific applications. SPR instruments marketed by Biacore have been widely used by sensor researchers around the world.

SPR measurements obtained with electrochemical reactions have been reported [12, 13]. An electrochemical reaction is a heterogeneous surface reaction. SPR instruments can also detect many types of electrochemically induced chemical reactions in surfaces modified with soft materials and without soft materials.

13.2.3 Electrochemistry

Metal electrodes are often modified with organic molecules and used to detect target materials. Thiol and its related materials can attach to a gold surface and form stable bonds. Alkanethiols and related materials are used to modify gold and other metal surfaces. One desirable feature of a monolayer on an electrode/substrate is its ability to limit the access of solution-phase molecules, both redox couples and electrolyte ions, to the electrode surface. A hydrocarbon layer in the monolayer can provide a substantial reduction in interfacial capacitance.

Self-assembled monolayers (SAMs) based on alkanethiols appear to survive a wide range of electrode potential and aqueous electrolyte compositions with no more than subtle changes in structure. In a strong base, the thiols can be desorbed quantitatively during cyclic voltammetry. A model adsorption and reductive desorption process is shown in Scheme 13.1.

When an alkanethiol SAM on an Au electrode is immersed in 0.5 M KOH and the potential is swept from 0 V to −1.5 V (versus Ag/AgCl), a reduction peak appears at −0.7 to −1.4 V (versus Ag/AgCl)] The peak position depends on the length of the alkane chain. Longer alkyl chains yield more negative peak potentials. The area of the peak is independent of the length of the alkane chain. The peak is assigned to the reductive desorption of the thiolate:

$$\mathrm{Au\text{ - }SC_nS + e^- + M^+ \rightarrow Au_0 + XC_nS^- M^+} \tag{13.5}$$

With an estimated surface roughness factor of 1.2, the peak area corresponds to coverage of 7.8 (±0.6) × 10^{-10} mol/cm^2, which is in good agreement with the coverage expected for a ($\sqrt{3}$ x $\sqrt{3}$)R30 overlay structure on Au(111) (7.7 × 10^{-10} mol/cm^2). The reproducibility of these results depends critically on the crystallinity of the gold electrode. The utility of the reductive stripping of thiols for surface coverage can be checked by forming a SAMs with both an electroactive and electroinactive thiol compounds. Cyclic voltammograms (CV) of ferrocene SAMs on Au(111) in $HClO_4$ electrolyte yield a coverage of 5.3 × 10^{-10} mol/cm^2 for elec-

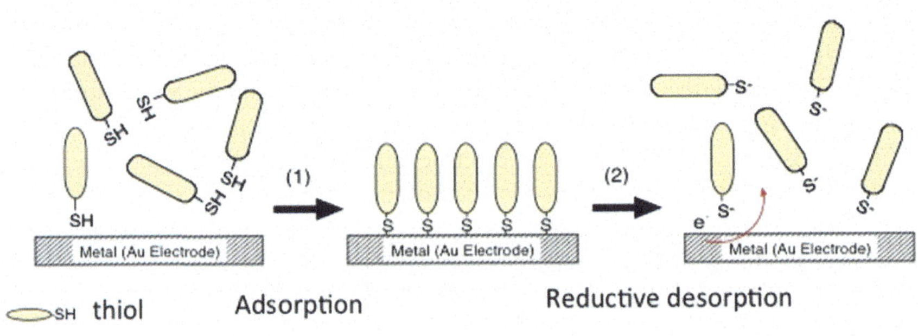

Scheme 13.1 Model of chemisorption (1) and electrochemical desorption (2) of thiol compounds on a metal (gold) electrode

troactive, ferrocene-terminated alkanethiol; $HS(CH_2)_{11}OOC\text{-}Fc$ [14]. The first value is somewhat higher than that obtained for a close-packed ferrocene layer, possibly indicating a degree of disorder in the monolayer.

Reductive desorption may also evaluate monolayer composition. With a mixed monolayer of short OH-terminated and COOH-terminated alkanethiols (for example, $HS(CH_2)_2OH$ and $HSCH_2COOH$), two reductive desorption peaks were observed. The two peaks correspond to the desorption of two alkanethiols [15]. Cathodic stripping peaks for SAMs formed from thiols ($CH_3(CH_2)_3SH$) and sulfides with an equivalent chain length ($CH_3(CH_2)_3\text{-}S\text{-}(CH_2)_3CH_3$) are identical in terms of peak position (−0.78 V vs. Ag/AgCl) and peak area (70 $\mu C/cm^2$). The cathodic stripping voltammogram for an SAM of ethyl phenyl sulfide shows two peaks of equal areas whose positions match the stripping peak positions for thiophenol and ethanethiol SAMs.

Short chain length alkanethiols are also oxidatively desorbed at +0.5 to +1.0 V (vs. Ag/AgCl in 0.5 M KOH) [16]. Approximately, three electrons are generated per one thiol. There were few reports for analogous stripping experiments at other metals.

$$Au\text{ - }SC_nX + 4OH^- \rightarrow Au^0 + XCnSO_2^- + 3e^- + 2H_2O \qquad (13.6)$$

The coverage of novel SAMs can be estimated by using reductive desorption results [17–20]. There should be a substantial charging current associated with the cathodic stripping current peak. As the monolayer is removed from the gold surface, the capacitance of the interface increases from about 1 to 20 $\mu F/cm^2$. In the CV experiment, the baseline charging current (equal to scan rate times capacitance) increases proportionally. The capacitive effect must be considered when evaluating the surface modification.

13.2.4 Field-Effect Transistors (FETs)

Field-effect transistors (FETs) are very useful for constructing bio-/chemical sensors because they can be made small by using current planar integrated circuit (IC) technology and have the advantage of a quick response time. FETs can measure the conductance of a semiconductor as a function of an electrical field perpendicular to a gate oxide surface. With a simple version, for example, a metal oxide semiconductor field-effect transistor (MOSFET), a p-type silicon substrate contains two n-type diffusion regions. The entire unit is covered with a silicon dioxide insulating layer on top of which a metal gate electrode is deposited (Fig. 13.5a). When a positive voltage is applied to the gate electrode, electrons are attracted to the surface. A conducting channel is created between the source and the drain, near the silicon dioxide interface. The conductivity of this channel can be modulated by adjusting the strength of the electrical field between the gate electrode and the silicon, perpendicular to the

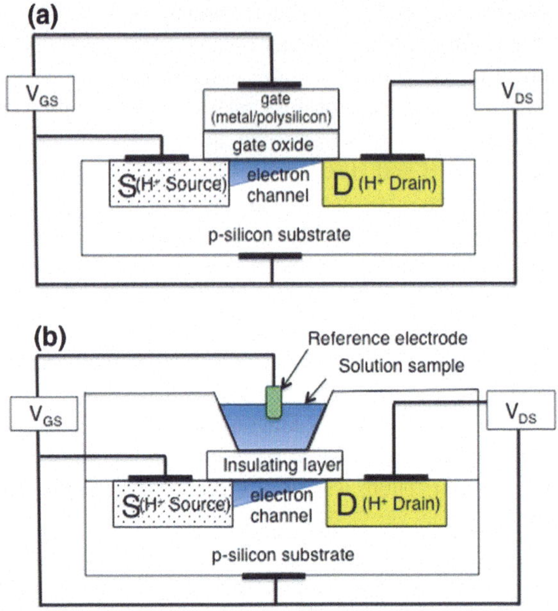

Fig. 13.5 Schematic diagrams of general **a** n-channel MOSFET and **b** ISFET. V_{GS} is the gate–source voltage, and V_{DS} is the drain–source voltage

substrate surface. At the same time, a voltage can be applied between the drain and the source, which produces a drain current between the n-regions. Another widely used type of integrated potentiometric sensor is based upon the ion-selective electrode (ISE), a relatively simple sensor that is widely used for the detection of ions in aqueous solutions. Figure 13.5 shows the general structure of an n-channel MOSFET and an ion-sensitive field-effect transistor (ISFET), which is the most common type of chemically sensitive field-effect transistor (ChemFET). In place of the gate and gate oxide of a conventional MOSFET, the ISFET has an ionic solution with a reference electrode immersed in the solution and an insulating layer appropriate for detecting a specific analyte. Many papers describing the development of ChemFET sensors have been published. These papers address structural issues, the selection of the insulation/detection layer, and the improvement of sensing accuracy. ChemFETs fabricated on a semiconducting substrate have the potential for integration with electronic circuits on the same chip. Because the current between their source and drain is a function of the surface density and type of charge of the adsorbed ions, ChemFETs, and other sensors are sensitive to pH. The gate structure of ChemFETs must be modified to improve performance or allow the measurement of specific chemicals.

13.3 Biosensing by Various Methods

13.3.1 QCM Measurements

There have been many previous reports on chemical/biosensing using the QCM method [1, 7]. Formerly, QCM in aqueous solution was used for such purposes as monitoring the construction of an alkanethiolate monolayer on a gold surface, and evaluating surface mass changes for electrochemical reactions [7]. After these early studies, the monolayer desorption from gold electrode surface was also monitored. The base–base (nucleic acid bases) interaction was detected in aqueous solution [21]. The layer-by-layer adsorption of biomolecules, polymers, and monolayers was also monitored by QCM for biosensing systems [1, 22].

With the new type of QCM measurements, resonance frequency and energy dissipation are measured simultaneously. In such measurements, the damping of the crystal oscillation (the dissipation factor) is measured in addition to the frequency shift, and it is also used to determine the adsorbed mass of thick and viscous layers. The classical handling methods for rigid films are not applicable to soft interfaces, soft materials, and soft films. Not only mass changes (m: ng/cm^2) but also density (ρ: kg/m^3), viscosity (η:kg/ms), elasticity (μ: Pa), and thickness (d: m) are needed for evaluating soft interfaces [2–6]. Applications that measured mass changes and other factors simultaneously have increased rapidly.

13.3.2 Surface Plasmon Resonance Measurements

SPR sensors have also been adopted to evaluate the affinities between (bio) molecules and surface modification with chemical materials. SPR biosensors have become tools for characterizing and quantifying biomolecular interactions. The number of publications reporting SPR biosensor applications for the detection of analytes (medical diagnostics, environmental monitoring, food analysis, and security) has been growing rapidly. SPR biosensors have become a dominant optical technology and have generated many papers, reports, and review articles [11].

With respect to SPR biosensors, an interesting monolayer is immobilized on the sensor surface. The surface chemistry has to be designed so that it can immobilize a sufficient number of biorecognition elements on the sensing surface while minimizing the nonspecific binding to the surface. Biorecognition elements need to be immobilized on the sensor surface without affecting their activity. Self-assembled monolayers (SAMs) of alkanethiolates have been widely used for the immobilization of biorecognition molecules on a gold surface. To obtain a sufficient number of biomolecular recognition elements and a nonfouling background, we need mixed SAMs with long alkylchains alkanethiolates terminated with recognition groups and we must avoid nonspecific adsorption groups, such as oligo(ethylene glycol)-terminated groups [15, 16].

We confirmed that tri(ethylene glycol)-terminated short alkanethiol is a tiny amphoteric modifier that includes a highly flexible—bulky—hydrophilic part and a short alkane backbone to provide dense packing. Tri(ethylene glycol)-terminated short alkanethiols work effectively to suppress nonspecific protein adsorption. The target molecules could be captured effectively when there is a co-adsorbed monomolecular layer on the substrate that uses this short amphoteric molecule and the target recognition molecule.

Figure 13.6a shows synthesized tri(ethylene glycol)-terminated alkanethiol for use as a protein repelling modifier and carbohydrate-alkanethiol (maltoside-terminated alkanethiol). Both chemicals can attach to the gold surface using thiol groups. We constructed a mixed monolayer on the gold surface for an SPR detection substrate. Figure 13.6c shows the specific recognition of Con A (Concanavalin A, lectin, protein) on the maltoside- and tri(ethylene glycol)-terminated alkanethiol mixed monolayer surfaces (Maltoside: tri(ethylene glycol) = 1:9). The nonspecific adsorption of Con A on the 100% tri(ethylene glycol) monolayer is also shown on the left of the figure (noise response). Con A recognized the α-1,4-bond of the maltoside part. Large signal responses caused by Con A adsorption were observed on each maltoside–ethylene glycol mixed interface. The signal-to-noise (S/N) ratio data are shown in Fig. 13.6d (signal response (S): specific recognition of Con A-maltoside on the mixed monolayer; noise response (N): nonspecific Con A adsorption on ethylene glycol terminated monolayer). The S/N ratio was greatly improved when much

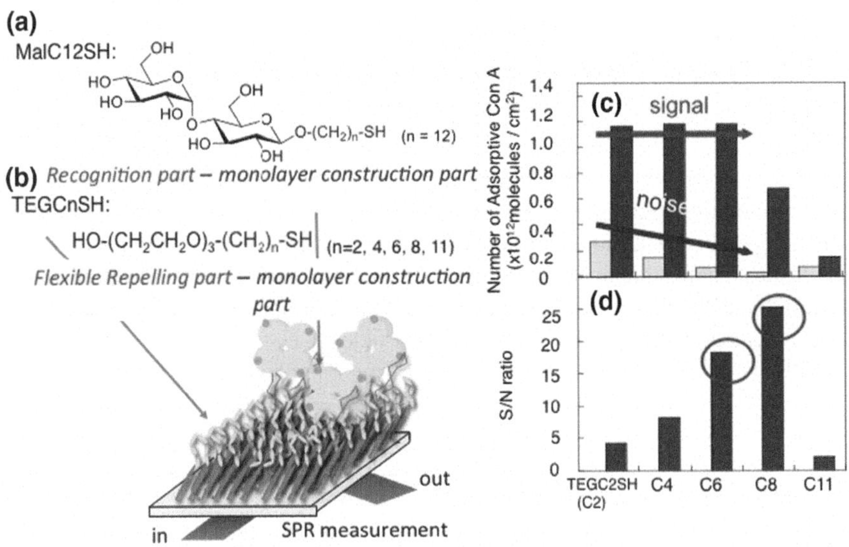

Fig. 13.6 Capturing Con A molecule (MalC12SH) and repelling molecules (TEGCnSH), and mixed Con A recognition molecular (**a**, **b**). Relationship between alkylchain length of TEGCnSH and number of adsorbed Con A (signal) (MalC12SH (10%)-TEGCnSH mixed monolayer) (**c**). Signal-to-noise ratio (signal response: noise response; S:N ratio) of Con A adsorption (**d**)

longer ethylene glycol molecules were used. The mixed monolayer model is shown in Fig. 13.6b. The combination of ethylene-glycol-terminated thiol with a longer alkylchain and a maltoside mixed layer was effective with regard to the specific recognition of lectin with a high S/N ratio.

13.3.3 Electrochemical Methods

Electrochemical methods, such as reductive desorption, are suitable ways to evaluate monolayer quality, behavior, and properties. Adsorbed alkanethiol molecules on gold and other metal surfaces were desorbed in an alkaline solution through a one-electron reductive path (RS-Au + e^- → RS- + Au). A model of the reductive desorption of thiolates from an Au surface is shown in Scheme 13.1. Figure 13.7 shows linear sweep voltammograms (LSVs) for the reductive desorption of our short ethylene-glycol-terminated alkanethiol monolayer on gold. The electrochemical desorption method has been widely used for the characterization of thiol monolayers, the total amounts of adsorbed molecules, the affinities of each thiolate to a gold surface, and the interaction between individual attached alkanethiol molecules, from the reductive desorption charges, reductive peak potentials, and peak configurations. The ethylene-glycol-terminated molecules were reduced at around −0.8 ~ −1.2 V (vs. Ag/AgCl

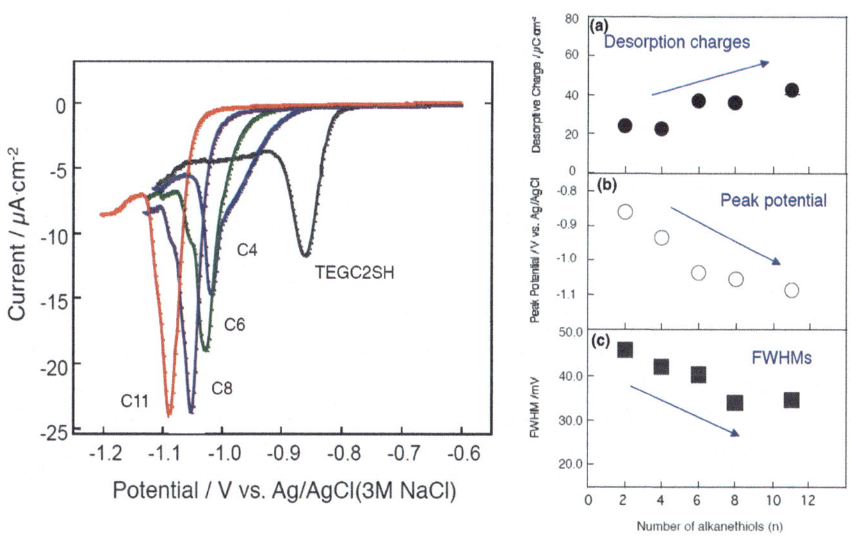

Fig. 13.7 (left) Linear sweep voltammograms (LSV) of a An(111) electrode modified with TEGC-nSH (n = 2~11) measured in 0.5 M KOH. Concentrations of TEGCnSHs solution are 1.0 μM (20% ethanol solution). Modification: 30 min. Sweep rate: 0.05 V/s. (right) Relationship between alkanethiol length and the electrochemical properties of reductive desorption peaks of TEGCnSH monolayers

(NaCl)) in 0.5 M KOH. The number of attached molecules is shown in the figure. The reductive desorption charge and surface concentration of the short alkylchain (C2: $(CH_2)_2$) monolayer were 23.6 $\mu C/cm^2$ and 1.5×10^{14} molecules/cm^2, respectively. On the other hand, with a longer alkylchain (C11: $(CH_2)_{11}$) monolayer, the charge and surface concentration were 42.3 $\mu C/cm^2$ and 2.6×10^{14} molecules/cm^2, respectively. The number of adsorbed shorter (C2) alkylchain molecules on gold was 45% less than the longer (C11) alkylchain molecules adsorbed on gold. Although the molecular density and homogeneity of adsorbed ethylene-glycol-terminated thiols might be lower than those of normal alkanethiol monolayers, the longer alkylchain monolayers were as stable as self-assembled monolayers of n-alkanethiols on gold.

Electrochemical measurements confirmed the crucial structural factors (packing density and durability) of ethylene-glycol-terminated alkanethiol monolayers in terms of excellent repellency. The monolayer acts as a nanobarrier despite the monolayer on the gold [23, 24].

13.3.4 Field-Effect Transistors (FETs)

The integration of biologically active materials together with an ion-selective field-effect transistor (ISFET) is one of the most attractive approaches. The ISFET has been introduced as the first miniaturized silicon-based chemical sensor. In spite of distinct difficulties with regard to practical applications, biologically modified field-effect transistors (BioFETs) have been developed. The BioFETs offer many advantages including small size and lightweight, good responses, high reliability, low output impedance, and the on-chip integration of biosensor arrays. BioFETs have a wide range of applications in medical, bioscience, industrial, and environmental monitoring for food, drugs, defense, an security.

BioFETs can be classified according to the biorecognition element that is used for detection: enzyme-modified FET (EnFET), immunologically modified EFT (ImmunoFET), DNA-modified FET (DNA-FET, or gene-modified FET (GenFET)), cell-potential FET (CPFET), and others. The principle of ISFETs and BioFETs has been reported in reviews [25, 26]. In recent years, many interesting experiments and results have accompanied the development of BioFETs and many applications have been reported [26–29].

Very important, popular, and highly sensitive sensing methods such as SPR, QCM, and electrochemical measurements (CV and others) employ the FET method. These sensing methods are very useful for evaluating the basic functions of a soft interface modified surface.

References

1. Marx KA (2003) Quartz Crystal microbalance: a useful tool for studying thin polymer films and complex biomolecular systems at the solution-surface interface. Biomacromolecules 4:1099–1120, and references are therein
2. Dahlin AB, Jønsson PM, Schmid E, Zhou Y, Hook F (2008) Synchronized quartz crystal microbalance and Nanoplasmonic sensing of biomolecular recognition reactions. ACS Nano 2:2174–2182
3. Muller MT, Yan X, Lee S, Perry SS, Spencer ND (2005) Lubrication properties of a brushlike copolymer as a function of the amount of solvent absorbed within the brush. Macromolecules 38:5706–5713
4. Irwin EF, Ho JE, Kane SR, Healy KE (2005) Analysis of interpenetrating polymer networks via quartz crystal microbalance with dissipation monitoring. Langmuir 21:5529–5536
5. Bonroy K, Friedt JM, Frederix F, Laureyn W, Langerock S, Campitelli A, Sara M, Borghs G, Goddeeris B, Declerck P (2004) Realization and characterization of porous gold for increased protein coverage on acoustic sensors. Anal Chem 76:4299–4306
6. Fogel R, Mashazi P, Nyokong T, Limson J (2007) Critical assessment of the quarts crystal microbalance with dissipation as an analytical tool for biosensor development and fundamental studies: metallophthalocyanine-glucose oxidase biocomposit sensors. Biosens Bioelectron 23:95–101
7. Shimazu K, Yagi I, Sato Y, Uosaki K (1992) In situ and dynamic monitoring of the self-assembling and redox processes of a ferrocenylundecanethiol monolayer by electrochemical quarts crystal microbalance. Langmuir 8:1385–1387
8. Shimazu K, Sato Y, Yagi I, Uosaki K (1994) Packing state and stability of self-assembled monolayers of 11-Ferrocenyl-1-undecanethiol on platinum electrodes. Bull Chem Soc Jpn 67:863–865
9. Shimazu K, Ye S, Sato Y, Uosaki K (1994) Simultaneous detection of structural change and mass transport accompanying the redox of a ferrocenylundecanethiol monolayer with the novel ft-ir reflection adsorption spectroscopy/electrochemical quartz crystal microbalance combined system. J Electroanal Chem 375:409–413
10. Sato Y, Mizutani F, Shimazu K, Ye S, Uosaki K (1997) Mass transport accompanied with electron transfer between the gold electrode modified with 11-Ferrocenylundecanethiol monolayer and redox species in solution—an electrochemical quartz crystal microbalance study. J Electroanal Chem 434:115–119
11. Homola J (2008) Surface plasmon resonance sensors for detection of chemical and biochemical species. Chem Rev 108:462–493 and references therein
12. Iwasaki Y, Horiuchi T, Niwa O (2001) Detection of electrochemical enzymatic reactions by surface plasmon resonance measurement. Anal Chem 73:1595–1598
13. Bin Y, Fu Q, Hou Y, Xiang J (2013) Real-time composition and deposition depth profile of metal alloys using electrochemical surface plasmon resonance. J Electroanal Chem 699:10–13
14. Walczak MM, Popenoe DD, Deinhammer RS, Lamp BD, Chung C, Porter MD (1991) Reductive desorption of alkanethiolate monolayers at gold: a measure of surface coverage. Langmuir 7:2687–2693
15. Weisshaar DE, Walczak MM, Porter MD (1993) Electrochemically induced transformations of monolayers formed by self-assembly of mercaptoethanol at gold. Langmuir 9:323–329
16. Widrig CA, Chung C, Porter MD (1991) The electrochemical desorption of n-alkanethiol monolayers from polycrystalline gold and silver electrodes. J Electroanal Chem 310:335–359
17. Sato Y, Mizutani F (2000) Electrochemical responses of cytochrome c on gold electrodes modified with nucleic acid base electrochimica acta 45:2869–2875
18. Sato Y, Ishikawa Y, Matsuura H, Uosaki K, Mizutani F, Niwa O (2005) Highly-sensitive detection of 6-mercaptopurine and its metabolites by electrochemical reductive desorption measurements. Electroanalysis 17:965–968

19. Imabayashi SI, Hobara D, Kakiuchi T (2001) Voltammetric detection of the surface diffusion of adsorbed thiolate molecules in artificially phase-separated binary self-assembled monolayers on a Au(111) surface. Langmuir 17:2560–2563
20. Hobara D, Ota M, Imabayashi SI, Niki K, Kakiuchi T (1998) Phase separation of binary self-assembled thiol monolayers composed of 1-Hexadecanethiol and 3-Mercaptopropionic Acid on Au(111) studies by scanning tunnering microscopy and cyclic voltammetry. J Electroanal Chem 444:113–119
21. Sato Y, Niwa O, Mizutani F (2007) Hydrogen bonding interaction between aminopurinethiol-monolayers and oligonucleotides by QCM and XPS measurements. Sensors Actuators B 121:214–218
22. Shi H, Song Z, Huang J, Yang Y, Zhao Z, Anzai JI, Osa T, Chen Q (2005) Effects of the type of polycation in the coating films prepared by a Layer-by-Layer desorption technique on the properties of amperometric choline sensors. Sens Actuators B 109:341–347
23. Sato Y, Yoshioka K, Tanaka M, Murakami T, Ishida NM, Niwa O (2008) Recognition of lectin with a high signal to noise ratio: carbohydrate-tri(ethylene glycol)-Alkanethiol Co-adsorbed monolayer. Chem Commun 4909–4911
24. Yoshioka K, Sato Y, Tanaka M, Murakami T, Niwa O (2010) Suppression of non-specific adsorption using densified tri(ethylene glycol) alkanethiols: monolayer characteristics evaluated by electrochemical measurements. Anal Sci 26:33–37
25. Joo S, Brown RB (2008) Chemical sensors its integrated electronics. Chem Rev 108:638–651
26. Schoning MJ, Poghossian A (2002) Recent advances in biologically sensitive field-effect transistors. Analyst 127:1137–1151 and references therein
27. Yumusak C, Singh TB, Sariciftci NS, Grote JG (2009) Bio-organic field effect transistors based on Crosslinked deoxyribonucleic Acid (DNA) gate dielectric app. Phys Lett 95:2633034
28. Goda T, Miyahara Y (2012) Interpretation of protein adsorption through its intrinsic electric charges: a comparative study using a field-effect transistor, surface plasmon resonance, and quartz crystal microbalance. Langmuir 28:14730–14738
29. Kim A, Ah CS, Park CW, Yang JH, Kim T, Ahn CG, Park SH, Sung GY (2010) Direct label-free electrical immunodetection in human serum using a flow-through-apparatus approach with integrated field-effect transistors. Biosens Bioelectron 25:1767–1773

Chapter 14
Nonprotein-Fouling, Hemocompatible, and Biospecific Surfaces Generated with Phospholipid Polymers

Yasuhiko Iwasaki

14.1 General Background of Protein Adsorption

The forces involved in protein adsorption onto surfaces are the competition between attractive interactions and nonspecific repulsions [1]. The major factors of attractive interactions are the entropic gain of water released due to hydrophobic interactions and the enthalpy loss due to the van der Waals and cation–anion attractive interactions. In general, the hydrophobic interaction operates over greater distances than does the van der Waals force [2].

Of the repulsive interactions, electrostatic repulsion is not significant in physiological conditions such as in body fluids or blood because the Debye length is usually less than 10 Å in physiological fluids. Repulsive hydration forces arise whenever water molecules bind to a surface containing hydrophilic groups [3]. Steric repulsion can provide long-range repulsion; the long-range aspect is provided by the thickness of the grafted layer of hydrophilic molecules. Thus, it is mainly long-range steric repulsion that can effectively overcome attractive interactions in the physiological environment.

To reduce protein adsorption on a polymer surface, some approaches for surface modification to induce repulsive interactions or lower attractive interactions have been taken (Table 14.1). Although the immobilization of hydrophilic polymers on material surfaces is effective in reducing nonspecific protein adsorption because retention of water is an important factor in repulsive interactions, hydrophilic nature is not enough to create satisfactory nonprotein-fouling surfaces. To control the nature of water bonded with a surface, the thickness of the hydrophilic polymer layers and the charge distribution of the surface might be important to create the ideal surface.

Y. Iwasaki (✉)
Department of Chemistry and Materials Engineering, Kansai University,
3-3-35 Yamate-Cho, Suita-Shi, Osaka 564-8680, Japan
e-mail: yasu.bmt@kansai-u.ac.jp

M. Maeda et al. (eds.), *Molecular Soft-Interface Science*,
https://doi.org/10.1007/978-4-431-56877-3_14

Table 14.1 Nonprotein-fouling surface compositions

Synthetic hydrophilic surfaces	Natural hydrophilic surfaces
PEG	Passivated proteins (e.g., albumin and casein)
Neutral polymers	Polysaccharides (e.g., hyaluronic acid)
Poly(2-hydroxyethylmethacrylate)	Liposaccharides
Polyacrylamide	Phospholipid bilayers
Poly(N-vinyl-2-pyrrolidone)	Glycoproteins (e.g., mucin)
Poly(N-isopropyl acrylamide) (below 31 °C)	
Anionic polymers	
Phosphorylcholine polymers	
Zwitterionic-functionalized materials	
Phosphorylcholine	
Sulfobetaine	
Carboxybetaine	

[2] Reproduced from Ratner B, Hoffman AS (2013) Non-fouling surfaces. In Ratner B, Hoffman AS, Schoen FJ, Lemons JE (eds) Biomaterials Science, 3rd edn. Copyright 2013 Elsevier Inc., Academic Press

14.2 Design of Phosphorylcholine-Bearing Polymers

What is the most appropriate surface for controlling interaction with proteins? Although this question has been considered for a long time in the creation of biomedical materials, the surface feature is still important issue because the most surface design is achieved by physicochemical approaches. On the other hand, a biomembrane surface might have the most suitable characteristics for controlling complex biological interactions. Recently, biomembrane structures have been adopted for preparing the surface of biomaterials. A model was proposed for designing blood-compatible polymeric materials with a cell membrane-like surface using a methacrylate monomer with a phosphorylcholine (PC) group and 2-methacryloyloxyethyl phosphorylcholine (MPC). Although research of the MPC polymer was first reported in 1978, the yield and purity might not have been sufficient to understand the properties of the MPC polymer [4]. After that, Ishihara et al. have established a refined and complete synthetic and purification process of MPC [5]. It has brought about excellent and considerable progress in the use of MPC polymers as biomaterials. Because MPC can readily be polymerized, numerous MPC polymers having a wide variety of molecular architectures, that is, random copolymers [6], block-type copolymers [7–10], graft-type copolymers [11–14], and terminal-functioned polymers [11, 15], have been prepared.

14.3 Mechanism of Nonprotein Fouling on MPC Polymer Surfaces

When protein molecules are adsorbed to a polymer surface, water molecules between the protein and the surface are displaced. A repulsive solvation interaction occurs whenever water molecules are associated with a surface that contains hydrophilic groups. Its strength depends on the energy necessary to disrupt the ordered water structure and ultimately dehydrate the surface. Even in the hydrophilic surface, the water structure has a major effect on protein adsorption and subsequent conformational change. Park et al. proposed a significant model for protein adsorption onto the surface of a polymer (Fig. 14.1). When proteins are adsorbed onto the surface of a polymer via hydrophobic interactions, an exchange of bound water between the protein and the surface is necessary [16]. The amount of bound water may therefore be the key parameter in understanding protein adsorption. Protein adsorption processes are considered to start with protein trapping by the random networks of water molecules on the material surface. The material surface, which cannot undergo hydrogen bonding with water, will then reduce protein adsorption. Kitano et al. showed that poly(MPC) (PMPC) chains did not significantly disturb the hydrogen bonds between their surrounding water molecules, suggesting that the phosphorylcholine groups may counteract electrostatic hydration [17, 18]. It has been considered that hydrated MPC polymers can accumulate a high free water fraction, and the nature of the water may be important for reducing protein adsorption on the surface. In contrast, the existence of nonfreezing water [19] or freezing bond water [20] in the hydrated MPC polymers and their origin has also been pointed out. Further, characterization of water at the surface of hydrated MPC polymers is needed to relate it with the nonprotein adsorption phenomena. It can be achieved by NMR relaxation time measurements or vibrational spectroscopy such as infrared, Raman, and sum-frequency generation (SFG), since these methods can probe the faster motion of water networks than thermal analysis and especially, vibrational spectroscopy can provide information on local water networks.

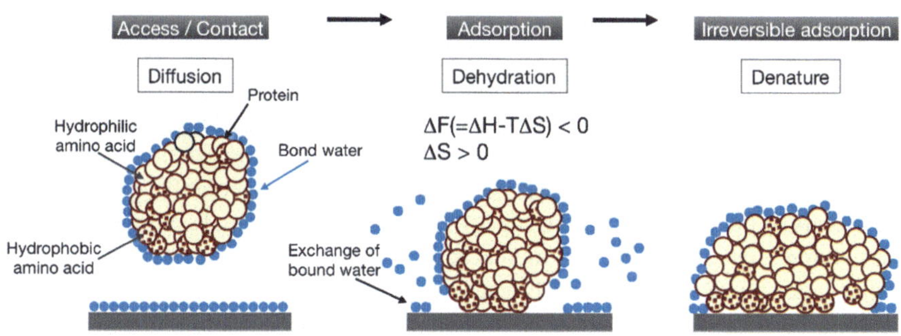

Fig. 14.1 Schematic diagram of protein adsorption on a polymer surface. Water molecules are removed from the sites of contacts between amino acid residues and a polymer surface

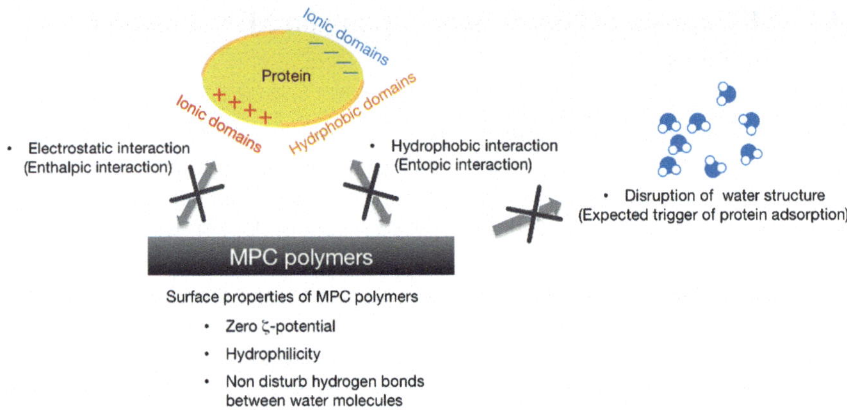

Fig. 14.2 Proposed mechanisms of nonprotein fouling on MPC polymer surface

The surface characteristics of an MPC polymer surface for the reduction of protein adsorption are summarized in Fig. 14.2. The phosphorylcholine group is very hydrophilic, which means it is unlikely to have hydrophobic interaction with proteins. Moreover, the phosphorylcholine polymer does not disintegrate the water structure around the polymer. This is a unique property of the phosphorylcholine polymer in comparison with conventional hydrophilic polymers. In addition, electrostatic interactions between a phosphorylcholine polymer surface and proteins are weak because the charge potential (ζ-potential) of phosphorylcholine is neutral [21]. Therefore, the phosphorylcholine polymer has several factors for reducing nonspecific protein adsorption. This polymer is one of the best polymer materials for constructing a nonbiofouling surface.

14.4 Hemocompatibility of MPC Polymers

The most common MPC polymer for coating onto a substrate is poly[MPC-co-*n*-butyl methacrylate (BMA)] (that is, PMB, Fig. 14.3a) with 30 mol% of the MPC unit in the polymer (PMB37) [5, 6, 22]. The MPC polymer coating surface shows a higher contact angle for water because the hydrophobic alkyl methacrylates enriched at the air–material interface reduce surface free energy. The surface is dynamic. Thus, surface equilibrium with aqueous media is necessary to orient the polar PC groups in the MPC units at the surface [23]. Figure 14.3b shows experimental columns containing polymer-coated poly(methyl methacrylate) beads and micrographs of the bead surface after contact with human blood without the use of an anticoagulant. Many adherent cells are observed on the hydrophobic poly(BMA) (PBMA). Furthermore, the activation and aggregation of the cells and clot formation were observed on the PBMA. In contrast, PMB37 effectively suppressed cell adhesion.

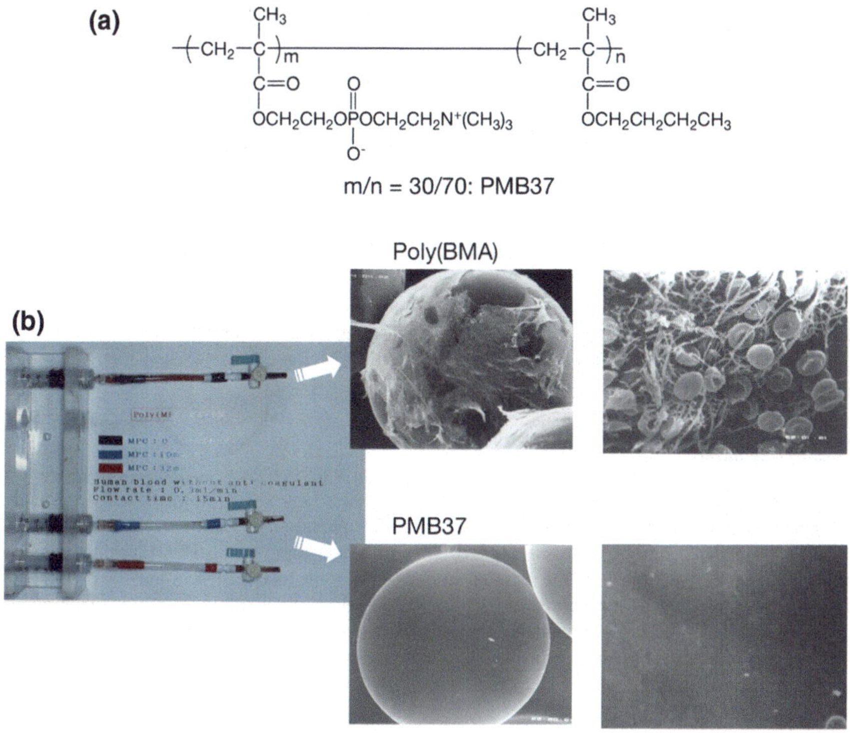

Fig. 14.3 **a** Chemical structure of PMB37. **b** Determination of blood compatibility of PMB37 by the microsphere column method

The hemocompatibility of MPC polymers has recently been determined in in vivo tests. As a typical experimental result, Soletti and co-workers have immobilized poly(MPC-co-methacrylic acid) (PMA73; Fig. 14.4a) onto the inner surface of electrospinning biodegradable poly(ester urethane)urea (ES-PEUU) tubes ($\phi = 1.3$ mm; Fig. 14.4b) and implanted them as aortic interposition grafts in a rat [24]. Patency at 8 weeks was 92% for the coated grafts compared to 40% for the non-coated grafts. Coating the inner surface of ES-PEUU tubes with PMA greatly impedes initial thrombus formation. Histology at 8 weeks demonstrated the formation of cellularized neo-tissue consisting of aligned collagen and elastin, as shown in Fig. 14.4c. Although elements of the regeneration process of neo-tissue have not yet been clarified, the novel coated grafts exhibited promising antithrombogenic and mechanical properties for small-diameter arterial revascularization.

MPC polymers have recently been applied to improve the blood compatibility of blood-contacting devices and cardiovascular devices [25] such as medical membranes for oxygenators and hemodialysis membranes, vascular prostheses, stents, cardiopulmonary bypasses, and ventricular assist devices.

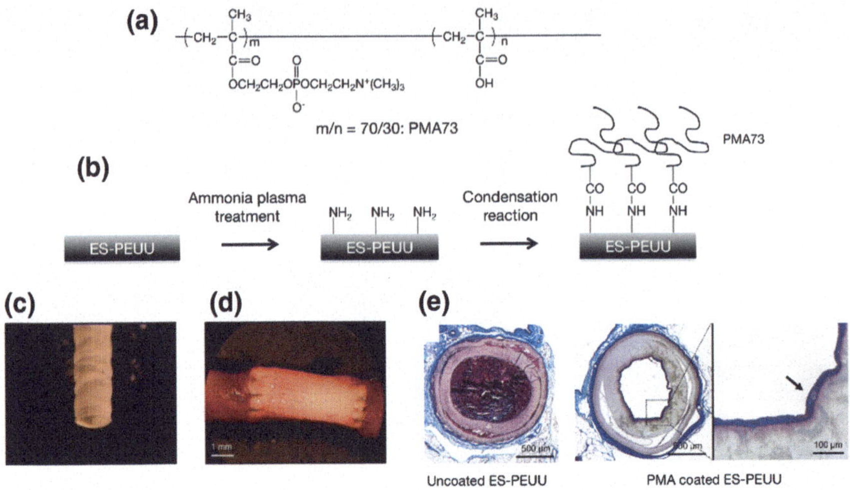

Fig. 14.4 **a** Chemical structure of PMA73. **b** Surface modification of ES-PEUU with PMA73. **c** Close-up of ES-PEUU tubing obtained at ×4. **d** Close-up of anastomosed ES-PEUU tubing obtained from a surgical microscope. Note the separate stitches made with 10-0 Prolene® and the proper size matching between the native rat aorta and the ES-PEUU tubing. **e** Histology (Masson's trichrome) following an 8-week period in vivo. Note the organized appearance of the neo-tissue in the lumen with a dense collagenous structure diffusively embedded with cells (arrow) [24] Reproduced with permission from Soletti et al. (2011) *J Biomed Mater Res A* 96: 436–448. Copyright 2011 Wiley-VCH

14.5 Regulation of Specific Molecular Interactions on MPC Polymer Surfaces

14.5.1 Protein-Conjugated Surfaces

Because MPC polymers can reduce nonspecific protein adsorption, as mentioned above, they contribute to generate surfaces that can regulate high-specific protein interactions. For protein conjugation with MPC polymers, copolymerization of MPC and a methacrylate having an active ester was recently performed. Ishihara et al. synthesized poly(MPC-co-BMA-co-*p*-nitrophenyloxycarbonyl poly(ethylene glycol) methacrylate (MEONP)) (PMBN, Fig. 14.5a) [26]. After the water-insoluble PMBN262 coats the substrate by solvent evaporation, the biomolecules can be immobilized. Nishizawa et al. immobilized antibodies on PMBN262 bearing active esters connected with different oxyethylene chains [27]. The MEONP units in PMBN262 could conjugate antibodies and enhance a specific signal. The specific signal was independent of the oxyethylene chain length. Furthermore, the activity of the antibodies immobilized on PMBN262 was preserved for more than 2 weeks (Fig. 14.5b).

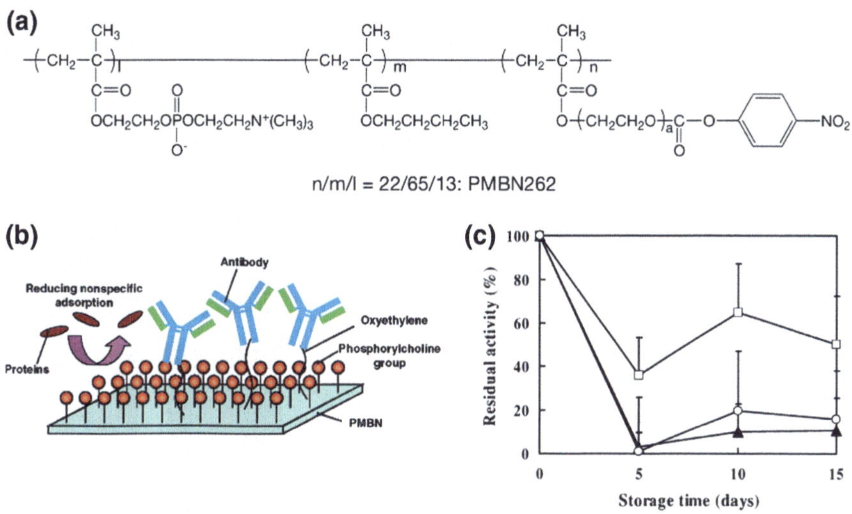

Fig. 14.5 **a** Chemical structure of PMBN62. **b** Schematic diagram of PMBN substrate for immunoassay. **c** Long-term stability of immobilized antibodies on various surfaces. Absorbance values (450–620 nm) obtained in ELISA were measured after the immobilized antibodies were stored at 37 °C under dry conditions. The residual activity was calculated as (absorbance values of human thyroid stimulating hormone (hTSH) = 10 μIU/mL) − (absorbance values of hTSH = 0 μIU/mL). The squares, circles, and triangles indicate the results with PMBN262 coating, no coating, and BSA blocking, respectively [27] Reproduced with permission from Nishizawa K et al. (2008) *Biomacromolecules* 9: 403–407. Copyright 2008 American Chemical Society

14.5.2 Protein-Imprinted Surfaces

MPC polymer surfaces are also useful as molecular imprinting (MIP) substrates. Fukazawa et al. prepared stamps bearing target proteins using PMBN-coated microparticles [28]. The particles were mixed with a photo-reactive MPC polymer (PMAz73, Fig. 14.6a) and SDS in aqueous solution and coated on quartz substrates, as shown in Fig. 14.6b. After UV irradiation, the particles were removed from the substrate by sonication. Protein recognition sites were constructed by integrating sodium dodecyl sulfate (SDS) as the ligand, which was immobilized with a biocompatible photo-reactive phospholipid polymer. When a BSA solution was added to a BSA-based MIP substrate, strong fluorescence was observed from the tryptophan residue of the BSA. In contrast, for an OVA-based MIP substrate and non-MIP substrate, no fluorescence was observed (Fig. 14.6c).

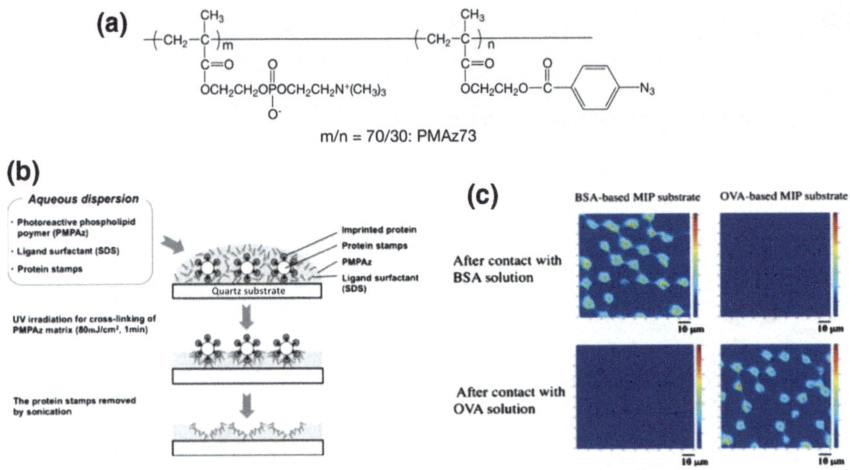

Fig. 14.6 **a** Chemical structure of PMAz37. Procedure for imprinting substrate using molecular integration. Fluorescence intensity images by UVFILM after contact with BSA or OVA solution to the BSA-based molecular imprinting (MIP) substrate and OVA-based substrate [28] Reproduced with permission from Fukazawa K et al. (2013) *Biosens Bioelectron* 40: 96–101. Copyright 2013 Elsevier Inc

14.5.3 Carbohydrate-Bearing Surfaces

Iwasaki et al. immobilized carbohydrate side chains on PMB37 for obtaining specific interaction toward cells [29]. They prepared poly(MPC-co-BMA-*co*-2-lactobionamidoethyl methacrylate (LAMA)) (PMBL, Fig. 14.7a) and coated it on substrates by solvent evaporation. Figure 14.7b shows the adsorption of lectins on the PMBL2 surfaces. Selective binding of a galactose-recognized lectin, RCA_{120}, to a PMBL2 surface was investigated by measurement of surface plasmon resonance. The binding of RCA_{120} to the PMBL surface was confirmed by a remarkable change in resonance angle. The apparent affinity constant of RCA_{120} to PMBL2 per LAMA unit was $2.77 \times 10^5\ M^{-1}$. When a glucose-recognized lectin, concanavalin A (ConA), was in contact with PMBL, no change in the resonance angle was observed and any nonspecific fouling of protein on PMBL was effectively reduced. The change in resonance angle ($\Delta\theta$) of various SPR sensor-coated polymers due to lectin binding is summarized in Fig. 14.7c. On PBMA and PBL1, $\Delta\theta$ was remarkable regardless of the type of lectin. This result indicates that nonspecific adsorption of lectin occurred on these surfaces. In contrast, $\Delta\theta$ was only detected on PMBL when the RCA_{120} was in contact with the surface and increased with an increase in the fraction of the LAMA unit in PMBL. Although absolute amount of RCA_{120} adsorbed on the PMBL1 surface was much lower than that on the PBL1 surface, nonspecific adsorption of ConA was completely reduced on the PMBL1 surface.

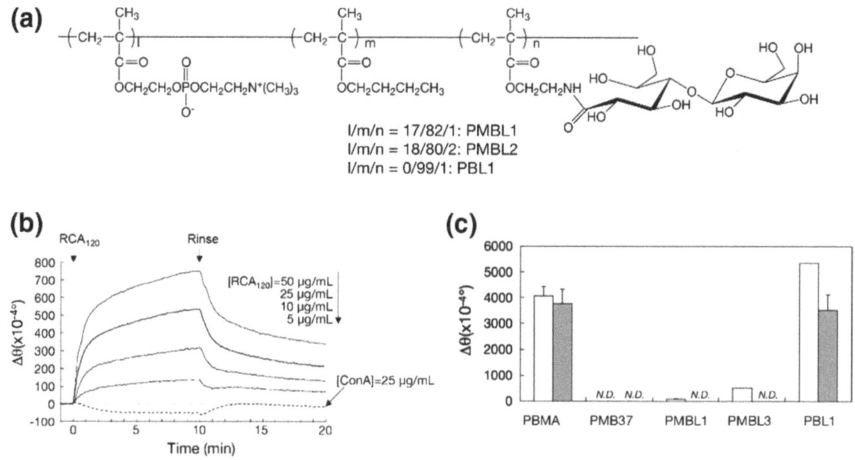

Fig. 14.7 **a** Chemical structure of PMBL and PBL. **b** SPR curve on PMBL3.0 after contact with RCA_{120} and ConA. **c** Change in resonance angle of polymer-coated SPR sensors after contact with lectin for 10 min: (white) RCA120 and (gray) Con A. [Protein] = 25 μg/mL [29] Reproduced with permission from Iwasaki Y et al. (2007) *Biomacromolecules* 8: 2788–2794. Copyright 2007 American Chemical Society

14.6 Conclusion

In this chapter, the nonprotein-fouling properties on MPC polymer surfaces were summarized. The highly hydration nature and zero ζ-potential of a phosphorylcholine group and its low interaction with water molecules are suitable for reducing protein adsorption and cell adhesion on the surfaces. MPC polymer surfaces then work as excellent coating materials to improve the hemocompatibility of medical devices and suitable platforms for the immobilization of any ligands, which interact with specific biomolecules and cells.

References

1. Lee JH, Li T, Park K (2001) Solvation interactions for protein adsorption to biomaterial surfaces. In: Mora M (ed) Water in biomaterials surface science. Wiley, Chichester
2. Ratner B, Hoffman AS (2013) Non-fouling surfaces. In: Ratner B, Hoffman AS, Schoen FJ, Lemons JE (eds) Biomaterials science, 3rd edn. Elsevier Inc., Academic Press, Waltham
3. Israelachvili JN (1985) Intermolecular and surface forces. Academic Press, New York
4. Kadoma Y, Nakabayashi N, Masuhara E, Yamauchi J (1978) Synthesis and hemolysis test of polymer containing phosphorylcholine groups. Koubunshi Ronbunshu 35:423–427
5. Ishihara K, Ueda T, Nakabayashi N (1990) Preparation of phospholipid polymers and their properties as polymer hydrogel membrane. Polym J 22:355–360

6. Ueda T, Oshida H, Kurita K, Ishihara K, Nakabayashi N (1992) Preparation of 2-methacryloyloxyethyl phosphorylcholine copolymers with alkyl methacrylates and their blood compatibility. Polym J 24:1259–1269
7. Inoue Y, Watanabe J, Takai M, Ishihara K (2004) Surface characteristics of block-type copolymer composed of semi-fluorinated and phospholipid segments synthesized by living radical polymerization. J Biomater Sci Polym Edn 15:1153–1166
8. Ma Y, Tang Y, Billingham NC, Armes SP, Lewis AL, Lloyd AW, Salvage JP (2003) Well-defined biocompatible block copolymers via atom transfer radical polymerization of 2-methacryloyloxyethyl phosphorylcholine in protic media. Macromolecules 36:3475–3484
9. Ma Y, Tang Y, Billingham NC, Armes SP, Lewis AL (2003) Synthesis of biocompatible, stimuli-responsive, physical gels based on ABA triblock copolymers. Biomacromolecules 4:864–868
10. Li Y, Armes SP, Jin X, Zhu S (2003) Direct synthesis of well-defined quaternized homopolymers and diblock copolymers via ATRP in protic media. Macromolecules 36:8268–8275
11. Ishihara K, Tsuji T, Kurosaki K, Nakabayashi N (1994) Hemocompatibility on graft copolymers composed of poly(2-methacryloyloxyethyl phosphorylcholine) side chain and poly(n-butyl methacrylate) backbone. J Biomed Mater Res 28:225–232
12. Iwasaki Y, Akiyoshi K (2004) Design of biodegradable amphiphilic polymers: well-defined amphiphilic polyphosphates with hydrophilic graft chains via ATRP. Macromolecules 37:7637–7642
13. Watanabe J, Eriguchi T, Ishihara K (2002) Stereocomplex formation by enantiomeric poly(lactic acid) graft-type phospholipid polymers for tissue engineering. Biomacromolecules 3:1109–1114
14. Watanabe J, Eriguchi T, Ishihara K (2002) Cell adhesion and morphology in porous scaffold based on enantiomeric poly(lactic acid) graft-type phospholipid polymers. Biomacromolecules 3:1375–1383
15. Samanta D, McRae S, Cooper B, Hu Y, Emrick T, Pratt J, Charles SA (2008) End-functionalized phosphorylcholine methacrylates and their use in protein conjugation. Biomacromolecules 9:2891–2897
16. Lu DR, Lee SJ, Park K (1991) Calculation of solvation interaction energies for protein adsorption on polymer surfaces. J Biomater Sci Polymer Edn 3:127–147
17. Kitano H, Sudo K, Ichikawa K, Ide M, Ishihara K (2000) Raman spectroscopic study on the structure of water in aqueous polyelectrolyte solutions. J Phys Chem B 104:11425–11429
18. Kitano H, Imai M, Mori T, Gemmei-Ide M, Yokoyama Y, Ishihara K (2003) Structure of water in the vicinity of phospholipid analogue copolymers as studied by vibrational spectroscopy. Langmuir 19:10260–10266
19. Morisaku T, Watanabe J, Konno T, Takai M, Ishihara K (2008) Hydration of phosphorylcholine groups attached to highly swollen polymer hydrogels studied by thermal analysis. Polymer 49:4652–4657
20. Hatakeyama T, Tanaka M, Hatakeyama H (2010) Studies on bound water restrained by poly(2-methacryloyloxyethyl phosphorylcholine): comparison with polysaccharide-water systems. Acta Biomater 6:2077–2082
21. Konno T, Kurita K, Iwasaki Y, Nakabayashi N, Ishihara K (2001) Preparation of nanoparticles composed with bioinspired 2-methacryloyloxyethyl phosphorylcholine polymer. Biomaterials 22:1883–1889
22. Iwasaki Y, Ishihara K (2005) Phosphorylcholine-containing polymers for biomedical applications. Anal Bioanal Chem 381:534–546
23. Yamasaki A, Imamura Y, Kurita K, Iwasaki Y, Nakabayashi N, Ishihara K (2003) The surface mobility of polymers having phosphorylcholine groups connected with various bridging units and their protein adsorption-resistance properties. Colloid Surf B 28:53–62
24. Soletti L, Nieponice A, Hong Y, Ye SH, Stankus JJ, Wagner WR, Vorp DA (2011) In vivo performance of a phospholipid-coated bioerodable elastomeric graft for small-diameter vascular applications. J Biomed Mater Res A 96:436–448
25. Iwasaki Y, Ishihara K (2012) Cell membrane-inspired phospholipid polymers for developing medical devices with excellent biointerfaces. Sci Technol Adv Mater 13:064101

26. Konno T, Watanabe J, Ishihara K (2004) Conjugation of enzymes on polymer nanoparticles covered with phosphorylcholine groups. Biomacromolecules 5:342–347
27. Nishizawa K, Konno T, Takai M, Ishihara K (2008) Bioconjugated phospholipid polymer biointerface for enzyme-linked immunosorbent assay. Biomacromolecules 9:403–407
28. Fukazawa K, Li Q, Seeger S, Ishihara K (2013) Direct observation of selective protein capturing on molecular imprinting substrates. Biosens Bioelectron 40:96–101
29. Iwasaki Y, Takami U, Shinohara Y, Kurita K, Akiyoshi K (2007) Selective biorecognition and preservation of cell function on carbohydrate-immobilized phosphorylcholine polymers. Biomacromolecules 8:2788–2794

Chapter 15
Stem Cell Purification on a Cell-Compatible, Cell-Specific Biointerface

Atsushi Mahara and Tetsuji Yamaoka

15.1 Introduction

Our bodies are composed of numerous different cell types. One fertilized ovum divides until maturation, and then organs are formed from these matured cells. Stem cells, which are present even in mature tissues, such as bone marrow and fat tissue, supply tissues and organs with progenitor and mature cells. Recently, stem cell-based tissue engineering has been investigated as a new therapeutic strategy [1] in which stem cell differentiation and proliferation are controlled to regenerate the structure and function of an organ based on cell engineering and biomaterial science techniques.

Research on stem cell biology has been rapidly developing over the past two decades. In 1998, Thomson et al. established embryonic stem (ES) cells from human blastocysts [2]. ES cells derived from early mammalian embryos are able to differentiate into all three embryonic germ layers, the endoderm, mesoderm, and ectoderm. Therefore, all functional tissues could be generated from an ES cell line. Although the differentiation ability of ES cells is attractive for medical applications, teratoma formation is a potential risk factor of ES cell transplantation therapy. In addition, the ethical problems associated with using human blastocysts have not yet been settled.

In 2006, Yamanaka and Takahashi reported the production of ES-like stem cells from adult fibroblastic cells by treatment with the four factors (Oct3/4, Sox2, c-Myc, and Klf4) [3]. These cells, called induced pluripotent stem (iPS) cells, share many similar features with ES cells, including differentiation ability, gene expression pattern, and teratoma formation. The ethical problems and immune rejection associated with ES cell use are not expected to be issues for iPS cell applications.

Mesenchymal stem cells (MSCs) are a class of adult progenitor cells that are capable of differentiating into several mesenchymal lineages. In 1999, Pittenger et al. first

A. Mahara · T. Yamaoka (✉)
Department of Biomedical Engineering, National Cerebral and Cardiovascular Center Research Institute (NCVC), Suita, Japan
e-mail: yamtet@ncvc.go.jp

M. Maeda et al. (eds.), *Molecular Soft-Interface Science*,
https://doi.org/10.1007/978-4-431-56877-3_15

reported the differentiation of marrow-derived MSCs into adipocytes, chondrocytes, and steoblasts [4]. MSCs are easily isolated from bone marrow and fat tissue by harvesting the cells that adhere to a tissue culture dish. The homogeneity of the isolated MSCs and their differentiation ability are still matters of debate; however, MSCs are the most extensively studied stem cells in the fields of biomaterials and tissue-regenerative medicine due to their ease of handling and their lack of an associated risk of teratoma formation.

In the field of stem cell research, the biointerface between the cell surface and the surrounding biomaterials has recently been attracting great attention. The molecular structure, mobility, and elasticity of the interface biomaterials have been shown to signal the cells to differentiate and proliferate. Engler et al. first reported that the elasticity of the matrix directs the differentiation lineage of stem cells [5] (Fig. 15.1). On soft surfaces with elasticities of 0.1–1 kPa, MSCs mainly differentiated into neuronal cells. In contrast, on hard surfaces with elasticities of 25–40 kPa, MSCs mainly differentiated into osteoblastic cells. Therefore, MSC differentiation is controlled

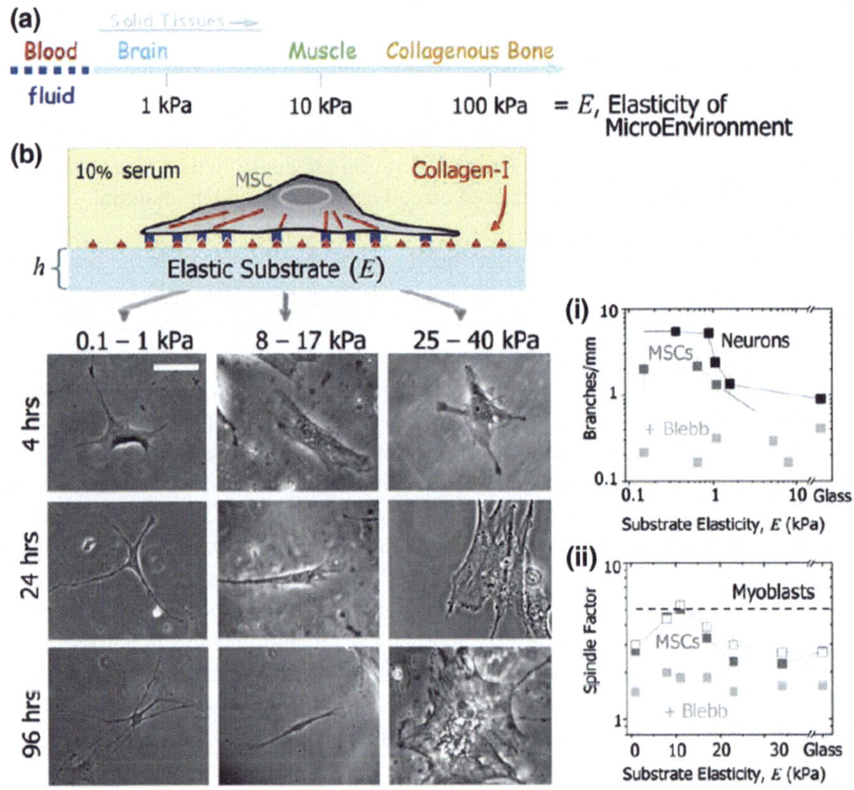

Fig. 15.1 Differentiation of MSCs on elasticity-controlled surfaces (Engler et al. Cell, 126, 677–689.)

not only by humoral factors but also by contact surface characteristics. Accordingly, the biointerface between the cells and the biomaterial surface is an important target for stem cell manipulation research.

ECM proteins are also known to influence stem cell differentiation in vivo and in vitro. Young et al. reported the changes in ECM composition that occur during cardiac differentiation, which was measured by RT-PCR. The relative expression levels of collagen in heart tissue increased, and the relative expression levels of fibronectin decreased as cardiac cells differentiated, indicating that fibronectin is preferable at the early stage of development [6]. It has also been shown that the cardiac stem cells in mature heart tissue are surrounded by a fibronectin-rich ECM called the cardiac niche [7]. We have also been studying the cardiac differentiation of MSCs and P19 embryonic carcinoma cells, one of the best model cell lines for cardiac research, and have determined the efficacy of various ECM compositions [8, 9].

One of the biggest reasons why the relationship between the biointerface and stem cell behavior has not been fully clarified despite such great efforts is the heterogeneity of these cell populations. Not all the stem cells treated with a differentiation–induction procedure differentiate into the target cells; instead, differentiation occurs stochastically. In addition, it is not easy to prepare homogeneous stem cells at a fully undifferentiated stage. Phinney et al. reported the heterogeneity of MSCs and differentiated cells [10]. Therefore, effective technologies to isolate and separate stem cells without reducing their multipotency are very important. Effective technologies for clinical stem cell therapy are also needed to secure their therapeutic benefits and safety.

15.2 Stem Cell Separation

As tissue stem cells mainly exist in the bone marrow, blood, and extracellular matrix, the process of cell separation can be divided into two steps. In the first step, the cells are separated from other components such as the ECM and soluble factors; for example, when the target cells exist in a tissue, the matrix is severed and digested with enzymes. Then, in the next step, the target cells are isolated. In the case of blood and bone marrow, the cells can be isolated by density-gradient centrifugation. Bone marrow-derived MSCs can be isolated by Ficoll centrifugation. In the second step, a single-cell population is isolated through the use of surface markers and/or the physical properties of the cells. This is the most important step for stem cell separation. Various separation methods have been reported that achieve a rapid and complete separation of homogeneous cells. These are summarized in Table 15.1.

Cell separation procedures are divided into two categories, physical and biological. In the physical and biological separation procedures, the cells are isolated by density and surface marker expression, respectively. In binary separation methods

Table 15.1 Cell separation methods

Category	Force used	Separation type	Method
Physical separation	Centrifugal force	Binary	Density-gradient centrifugation
	Adhesive force	Binary	Adhesion cell culture
Biological separation	Electrical force	Continuous Binary	DEP (Dielectrophoresis) FACS (Fluorescence-activated cell sorting)
	Magnetic force	Continuous Binary	DMFF (Dipole magnet flow fraction) MACS (Magnetic-activated cell sorting)
	Hydrodynamic force	Continuous Binary	Column Membrane (antibody-immobilized mesh)

that use a cell-specific antibody, the cells are separated into two groups based on positive and negative surface marker detection. Fluorescence-activated cell sorting (FACS) [11] and magnetic-activated cell sorting (MACS) [12] are well-known traditional cell separation methods based on surface marker labeling with fluorescently and magnetically labeled antibodies, respectively. MACS is a very simple and convenient method; however, it only allows for binary separation. When an antibody for a specific cell population is available, the target cell population is easily separated using these methods. Density-gradient centrifugation and cell culture plate adhesion are more simple ways to isolate cells. However, since the cell populations are not separated based on their biological features, the harvested cells are heterogeneous from a biological point of view. In continuous separation, a heterogeneous cell suspension can be separated into multiple outlet fractions based on continuously changing cell factors, such as cell size, surface charge, or surface marker density. Specific cell surface markers are particularly important because they are relevant to the differentiation stage and phenotype. Progenitor and stem cells are defined by their specific marker expression pattern [13]. Therefore, continuous separation of stem cells based on surface marker expression pattern and/or density would be a promising method for the purification of immature cells and differentiated cells. To this end, the continuously changing features of cells should be converted into some detectable signal. In the FACS system, cell surface marker density is converted into fluorescence intensity, and cells with a given marker molecule density can be purified by setting the gate on the fluorescence intensity. However, the very strong shear stress during FACS purification severely affects cell viability, and the purified cells carry a fluorescently labeled antibody. To overcome the limitations associated with current cell separation methods, we have been developing a novel, label-free continuous stem cell separation system that utilizes a cell surface marker-specific biointerface. In this system, surface marker density is converted to cell rolling velocity on the biointerface.

15.3 Cell Rolling Mechanisms

Cell rolling phenomenon is attractive for the cell separation because the rolling velocity depends on the interaction point density on the biointerface-immobilized ligand and its density. Leukocyte rolling is the most well-known cell rolling. When the blood vessel receives a signal from the inflammatory site, the endothelial cells expressed the selection. In this step, rolling velocity of the leukocyte is gradually reduced through the interaction between the expressed selectin and leukocyte surface receptor. Finally, the leukocyte is trapped and moved to the inflammatory region throughout the luminal layer. The rolling velocity is strictly controlled by the ligand and surface marker of cells. Hammer et al. theoretically evaluated the dynamic cell motion for receptor-mediated cell adhesion [14]. Andrian et al. revealed the leukocyte rolling mechanisms as the temporal interaction between the leukocyte surface marker and endothelial cells on the luminal surface [15]. The rolling mechanisms have been also reported by many groups [16]. This specific interaction of cells has been applied in drug screening, local delivery of therapeutics [17], and studies of cell regulation [18]. That is, cell rolling velocity is correlated with the surface maker and expression level, and marker expression level would be recognized by the cell rolling velocity.

In the recent our research, the cell rolling has been focused as the mechanisms for the stem cell separation [19]. When the cells were rolled on the ligand-immobilized surface, the rolling velocity was restricted by the temporal interaction between the cell surface marker and immobilized ligand (Fig. 15.2). Moreover, the rolling velocity was depended on the marker expression level in the theoretical study, so we assumed that the specific cell population can be isolated on the ligand-immobilized biointerface by the rolling velocity without any cell-labeling.

15.4 Cell Separation on Ligand-Immobilized Column

To evaluate the cell separation ability on ligand-immobilized surface, anti-CD34 antibody-immobilized polyethylene (PE) tube (inner diameter = 1 mm) was constructed as cell separation column [19]. Ozone-induced graft polymerization was carried out to prepare antibody-immobilized surface. Polyacrylic acid was grafted on the luminal surface of PE tube, and the antibody was reacted with the amino groups by water-solved carbodiimide activation method. The immobilized antibody was approximately 180 $\mu g/m^2$ under the suitable reaction condition. The density of

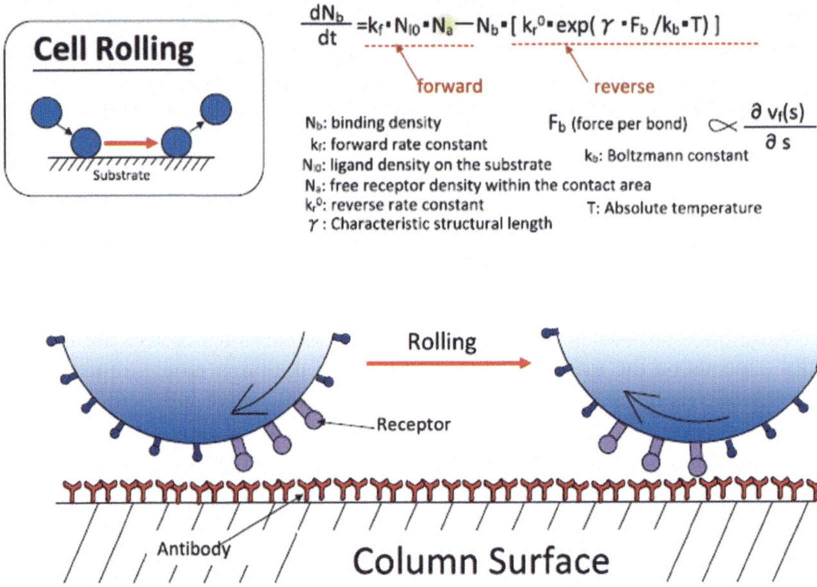

Fig. 15.2 Cell rolling mechanisms on ligand-immobilized surface

the antibody was calculated as 1.2×10^9 mol/m^2. Greenberg et al. reported that the 10^9 mol/m^2 (800 sites/μm^2) was suitable for cell rolling on the ligand-immobilized substrate in an experimental as well as theoretical study [20]. The ligand-immobilized tube was connected to the syringe pump, and the cell separation system was constructed (Fig. 15.3a). The column was fixed at a tilt of 20–90°, and elution yield of the injected KG-1a (CD34 positive) cells was evaluated under 50 μL/min of media flow. When the tilt angle was 90°, 80–100% of injected cells were eluted from the column. On the other hand, the 50% of cells were eluted at the tilt of 20°. These results indicated that injected cells were scarcely interacted with the immobilized antibody at a tilt of 90°, and 50% of cells were trapped on the immobilized surface at a tilt of 20°. On the interface between the cells and the ligand surface at 20° of tilt angle, the cells were attached on the surface through multivalent interaction, and the cells were not rolled on the surface under the media flow. On the other hand, when the media flow was changed to 1 mL/min, injected cells were completely eluted at any tilt angle. Consequently, media flow and tilt angle were largely affected by the interaction of the cells on the ligand. Finally, 50 and 600 μL/min of media flow were selected as the injection media flow and separation media flow. KG-1a (CD34 positive) cells and HL-60 (CD34 negative cells) were used for the evaluation of cell rolling column. Figure 15.4 shown a data of the separation pattern of KG-1a and HL-60 cells on anti CD34 antibody-immobilized and unmodified column . When the KG-1a cells were

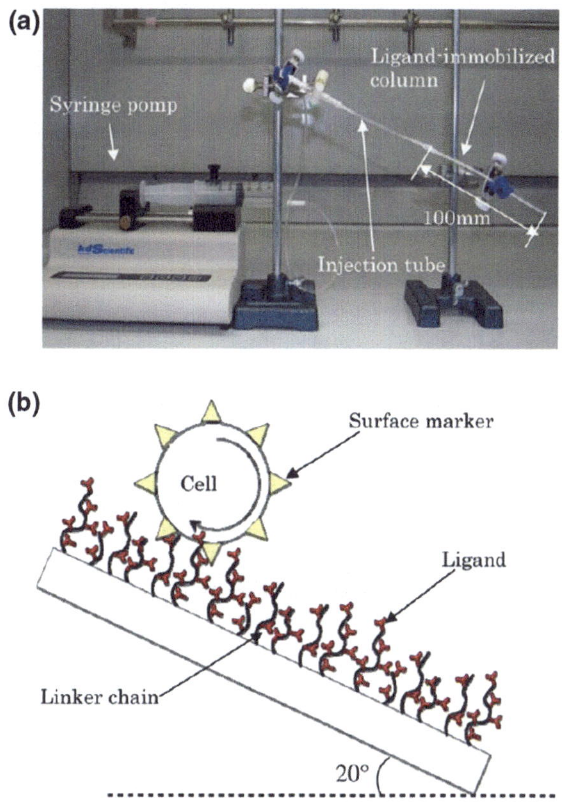

Fig. 15.3 **a** Cell rolling column and **b** cell separation mechanisms on ligand-immobilized biointerface

injected into the antibody-immobilized column, two peaks were observed around the fraction 2 and 7. On the other hand, the single peak was observed in another control conditions. In the first fraction observed around fraction 2, the cells would be eluted without any interaction on the surface. That is, the cells eluted in around the second peak were interacted with the immobilized antibody, and the cells with high marker density contain the latter fraction. When the rolling velocity was evaluated using antibody-immobilized microchip, the rolling velocities of KG-1a and HL-60 cells were observed as 0.45 and 0.6 mm/s, respectively. These experimental data indicated that the cells could be separated from the ligand-immobilized interface through the interaction of cell surface and immobilized ligand. Since the established homogeneous cell lines were selected, the separation pattern was not largely distributed. In the case of the stem cells, the marker expression and its density would be distributed in a wide range.

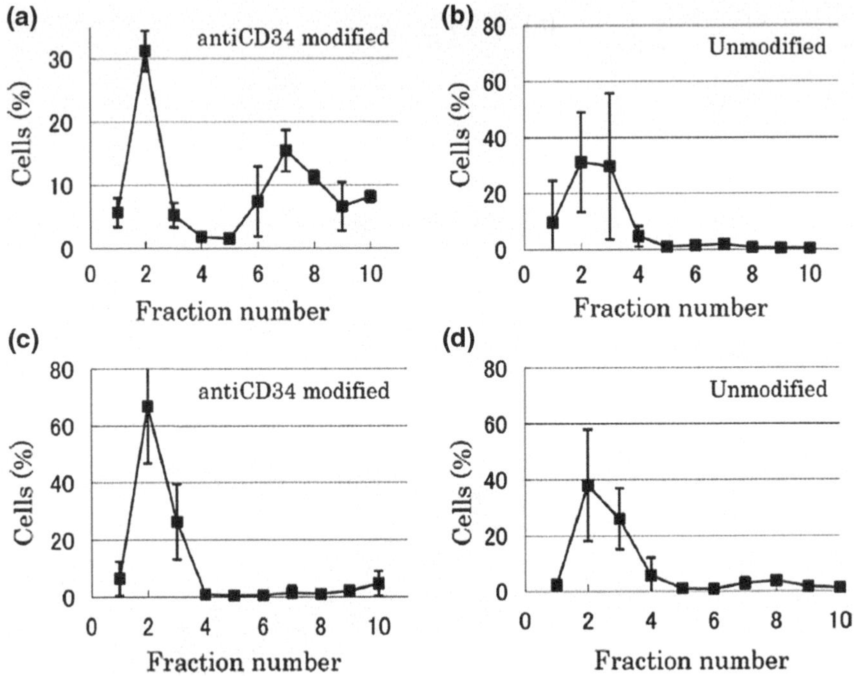

Fig. 15.4 Elution profile for KG-1a (**a** and **b**) and HL-60 cells (**c** and **d**) on the anti-CD34 modified column (**a** and **c**) and unmodified column (**b** and **d**)

15.5 Mesenchymal Stem Cells Separation

Mesenchymal stem cells (MSCs) are tissue-derived stem cells capable of differentiation to mesenchymal lineage. Surface marker profile such as marker density and its variation changes with the differentiation and development. The marker expression pattern of murine MSCs significantly differs from the passage level [21]. Consequently, isolation of homogeneous population of MSCs is important to define the cellular characteristics of the cells. To assess the usability of the cell rolling column, the separation efficiency and differentiation characteristics of isolated MSCs were evaluated [22]. MSCs were isolated from murine bone marrow as the adherent cells population in which isolation method was known as gold standard methods [4]. Morphology of the crude MSCs and surface marker expression is shown in Fig. 15.5. The shape of the MSCs was shown as a fibroblast-like cells, and expressions of CD29, 44, 81, and Sca-1 were confirmed. Expression of CD34 was also confirmed in murine crude MSCs, and this was also reported by Umezawa et al. and some group [23]. CD34 is a progenitor or stem cell marker of the hematopoietic lineage, and the expression continuously decreased with the culture period. Therefore, CD34 expression would be closely related to the differentiation stage of the cells. Hence, the anti-CD34 antibody was selected as the immobilized ligand for the column of MSCs separation. Separation condition determined by the previous study was employed for the MSCs separation. To prevent the elution of the cell without any interaction on the column surface, 0.5 mm of the inner diameter of the column was used. Separation pattern of the crude MSCs on anti-CD34 antibody-immobilized column and unmodified column is shown in Fig. 15.6. When the crude MSCs were separated on the anti-CD34 antibody-immobilized column, two peaks were observed around fractions 3 and 6. The isolated MSCs were isolated with the marker expression density confirmed by FACS analysis. To evaluate the differentiation ability of isolated MSCs, the cells were cultured in the osteoblastic differentiation medium for 7 days. The differentiation was evaluated by the calcium deposition level stained with alizarine red S (Fig. 15.7). The isolated MSCs in fractions 3 and 6 were strongly stained with the alizarine red S. These data indicated that two populations with high differentiation ability for osteoblast were contained in the crude MSCs. The differentiation stage was also evaluated using the real-time PCR experiments. CBFA1 is a key factor for mature osteoblastic differentiation, and the expression is necessary for calcium deposition on the cells [24]. On the results of PCR experiments, the expression level of CBFA1 in fraction 3 and 6 was approximately 10-fold higher than other fraction. It was reported that osteoblastic progenitor cells were enriched in CD34-positive population from bone marrow [25]. Based on these reports and CD34 expression level of the isolated MSCs, MSCs with high differentiation ability were contained in early fraction, and a progenitor-like cell was enriched in the fraction 6. These finding showed that the ligand-immobilized column could separate homogeneous cell populations by cell rolling mechanisms and some cell populations contained in the crude MSCs.

15.6 Conclusion

Biointerface between the cells and material surface controls the cell behavior such as proliferation and differentiation. Surface molecules strictly recognize the surrounding environment through the specific receptor–ligand interaction. However, the detailed molecular mechanisms of cell recognition to the surface is not completely clarified. On the other hand, cell rolling is induced by the temporary interaction between the cell surface and ligand, and the velocity is regulated by the ligand–receptor interaction in the surface. In our research, new cell separation system could be developed using the cell rolling mechanisms. The cell separation was successfully achieved by the specific interaction. In near future, the cell behavior and separation would be strictly regulated based on the molecular characteristics of biointerface.

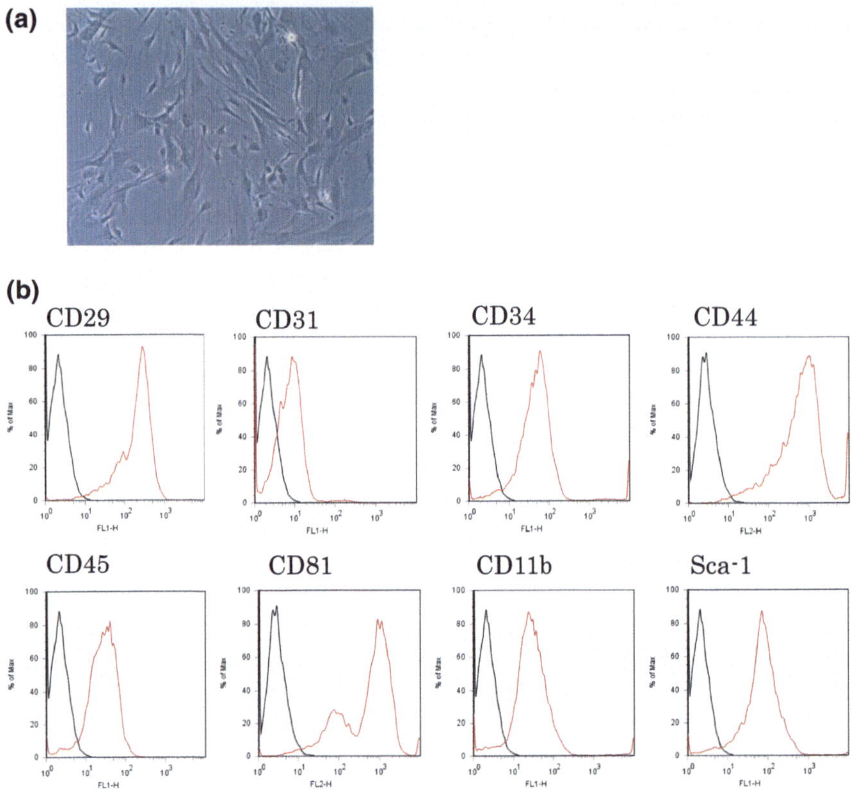

Fig. 15.5 **a** Morphology of murine MSCs and **b** surface marker expression of the crude MSCs evaluated by FACS analysis

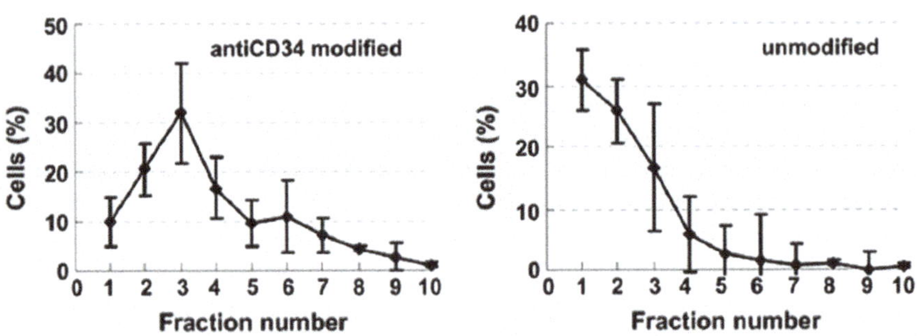

Fig. 15.6 Separation profile of murine crude MSCs on the anti-CD34 antibody-immobilized column or unmodified column

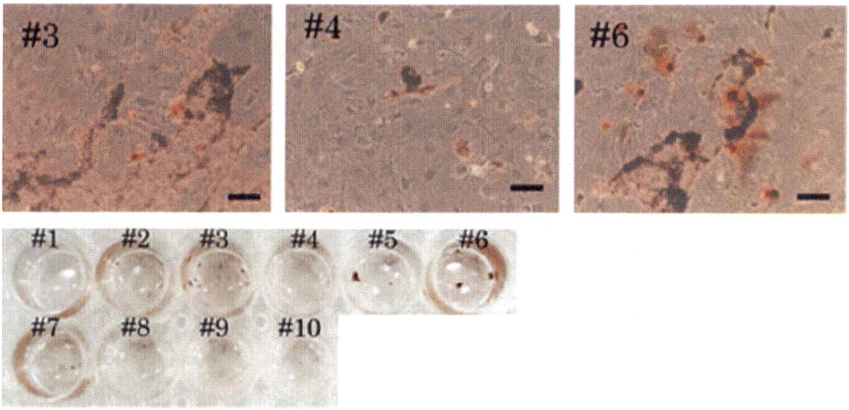

Fig. 15.7 Alizaline red S staining of isolated MSCs on cell rolling column. The isolated MSCs were culture with osteoblastic medium for 7 days

References

1. Langer R, Vacanti JP (1993) Tissue engineering. Science 260:920–926
2. Thomson JA, Itskovitz-Eldor J et al (1998) Embryonic stem cell lines derived from human blastocysts. Science 282:1145–1147
3. Takahashi K, Yamanaka S (2006) Induction of pluripotent stem cells from mouse embryonic and adult fibroblast cultures by defined factors. Cell 126:633–676
4. Pittenger MF, Mackay AM et al (1999) Multilineage potential of adult human mesenchymal stem cells. Science 284:143–147
5. Engler AJ, Sen S et al (2006) Matrix elasticity directs stem cell lineage specification. Cell 126:677–689
6. Young JL, Engler AJ (2011) Hydrogels with time-dependent material properties enhance cardiomyocyte differentiation in vitro. Biomaterials 32:1002–1009
7. Bearzi C, Rota M et al (2007) Human cardiac stem cells Proc Natl Acad Sci U S A. 104:14068–14073
8. Miskon A, Mahara A et al (2010) A suspension induction for myocardial differentiation of rat mesenchymal stem cells on various ECM proteins Tissue Eng Part C 16:979–987
9. Miskon A, Ehashi T et al (2009) Beating behavior of primary neonatal cardiomyocytes and cardiac-differentiated P19CL6 cells on different extramatrix components. J Art Org 12:111–117
10. Phinney DG, Hill K et al (2006) Biological activities encoded by the murine mesenchymal stem cell transcriptome provide a basis for their developmental potential and broad therapeutic efficacy StemCells 24:186–198
11. Fulwyler (1965) Electronic separatin of biological cells by volume Science 150:910–911
12. Molday RS, Yen SPS, Rembarum A (1977) Application of magnetic microspheres in labeling and separation of cells. Nature 268:437–438
13. Osawa M, Hanada K et al (1996) Long-term lymphohematopoietic reconstituteion by a single CD34-low/negative hematopoietic stem cell Science 273:242–245
14. Hammer DA, Lauffenburger DA (1987) A dynamic model for receptor-mediated cell adhesion to surfaces. Biophys J 52:475–487
15. Von Andrian UH, Chambers JD et al (1991) Two-step model of leukocyte-endothelial cell interaction in inflammation: distinct roles for LECAM-1 and the leukocyte beta 2 integrins ni vivo Pro Natl Acad Sci USA 88:7538–7542

16. Hanley WD, Wirtz D, Konstantopoulos K (2004) Distinct kinetic and mechanical properties govern selectin-leukocyte interactions. J Cell Sci 117:2503–2511
17. Omolola EA, Hammer DA (2005) vitro characterization of leukocyte mimetic for targeting therapiutics to the endothelium using two receptors. Biomaterials 26:7136–7144
18. Greenberg AW, Hammer DA (2001) Cell separaion mediated by differential rolling adhesion. Biotechnol Bioeng 73:111–124
19. Mahara A, Yamaoka T (2010) Antibody-immobilized column for quick cell separation based on cell rolling. Biotechnol Prog 26:441–447
20. Greenberg AW, Kerr WG, Hammer DA (2000) Relationship between selectin-mediated rolling of hematopoietic stem and progenitor cells and progression in hematopoietic development. Blood 95:478–486
21. Wieczoreck G, Steinhoff C et al (2003) Gene expression profile of mouse bone marrow stromal cells determined by cDNA microarray analysis. Cell Tissue Res 311:227–237
22. Mahara A, Yamaoka T (2010) Continuous separation of cells of high osteoblastic differentiation potential from mesenchymal stem cells on an antibody-immobilized column. Biomaterials 31:4231–4237
23. Gojo S, Gojo N et al (2003) vivo cardiovasculogenesis by direct injection of isolated adult mesenchymal stem cells. Exp Cell Res 288:51–59
24. Komori T, Yagi H et al (1997) Targeted disruption of cbfa1 results in a complete lack of bone formation owing to maturational arrest of osteoblasts Cell 89:755–764
25. Chen JL, Hunt P et al (1987) Osteoblast precursor cells are found in CD34 + cells from human bone marrow Stem Cells 15:368–377

The manufacturer's authorised representative in the EU is Springer Nature Customer Service Centre GmbH, Europaplatz 3, 69115 Heidelberg, Germany. If you have any concerns regarding our products, please contact ProductSafety@springernature.com

Printed and bound by CPI Group (UK) Ltd, Croydon, CR0 4YY

29/07/2026

02180629-0001